George Apostol

Chromatographic Methods

Editors:
W. Bertsch
W.G. Jennings
R.E. Kaiser

Rudolf E. Kaiser · Anthony J. Rackstraw
Computer Chromatography

Computer Chromatography

Volume 1

by
Rudolf E. Kaiser and
Anthony J. Rackstraw

with 43 Figures and 11 Tables

Dr. Alfred Hüthig Verlag
Heidelberg · Basel · New York

Dr. rer. nat. Rudolf E. Kaiser
Institut für Chromatographie
P.O. Box 1141
D-6702 Bad Dürkheim 1

Anthony J. Rackstraw
Dr. Alfred Hüthig Verlag GmbH.
P.O. Box 102869
D-6900 Heidelberg

CIP-Kurztitel der Deutschen Bibliothek

Kaiser, Rudolf E:
Computer chromatography / by Rudolf E. Kaiser
and Anthony J. Rackstraw. -
Heidelberg; Basel; New York:
Hüthig (Chromatographic methods)
NE: Rackstraw, Anthony
Vol. 1 (1983).
ISBN 3-7785-0885-7

Printed in Germany
Written with Apple Writer II
on an Apple II computer at IfC,
Bad Dürkheim, and Dr. Alfred Hüthig Verlag, Heidelberg
Transferred from disk to a
Compugraphic TW 7700 Photosetter by
Druckerei Bitsch KG · Birkenau
Printing: Druckerei Bitsch KG · Birkenau

Preface

This book is about an on-going revolution—about the liberation or reliberation of the analyst from the fetters of his own inflexible thought processes and from the power of the omnipresent black box. The reader may greet some of the statements made within its covers with incredulity. The regeneration of raw data by means of cubic spline functions and the real possibility of subtracting the baseline from a thin-layer chromatogram are just two far-reaching claims which will not fail to cause many raised eyebrows. However, all these claims can be tested with the aid of a tool now within easy reach of almost every analyst, *i.e.* a microcomputer with a minimum of peripheral equipment. The open-minded analyst will then soon discover the truth of many an "iconoclastic" pronouncement to be found in this volume. This very book is also a tribute to that device which has just been proclaimed "Machine of the Year" by Time Magazine—the computer; at least in the opinion of the authors.

One of us (*REK*) wishes to express his gratitude to many colleagues from the early days of mainframes to the micros of today. To mention just a few: *H. Kienitz, G. Guenzler, F. Caesar, H. Fiedler, Margit Haferkamp, W. Stoll, G. Schomburg, E. Ziegler, D. Henneberg, D.A. Deans, L. Rohrschneider, S. Ebel, H. Knoesel, E. Wagner, H. Ostermann P.K. Warme, S. Cram, W. Bertsch, W. Jennings, V. Pretorius, T.S. Buys, G. Suess, Edit Vogel, M. Prosek, A. Medja, A. Miller,* and last but not least *R. Rieder.* We thank *Katharina Radvakova* for her help with the computer manuscript.

Anthony J. Rackstraw, Heidelberg
Rudolf E. Kaiser , Bad Duerkheim

IfC, May 1983

Contents

1 Introduction

The present Volume 1 of "Computer Chromatography" will NOT discuss big computer solutions in chromatography—nor will the subsequent volumes. Instead, the whole series will feature exclusively small, dedicated computers, *i.e.* dedicated to a single chromatograph. This may strike the reader as being at complete variance with modern practice, for it is nowadays not uncommon in large laboratories to have 40...80 or even more chromatographs on-line to one more or less "big" computer. Having worked with these big systems for years the author simply can say that they do not allow "COMPUTER CHROMATOGRAPHY", they just computerize chromatographs.

Computer chromatography is more. Of course, it would be technically feasible to make big computers offering "computer chromatography". And to avoid misunderstanding: if a computer is big (gigabyte programs, gigabyte memory), AND is on-line with chromatographs, AND offers immediate access to computer help, then such a system is just as good as single dedication, provided that it is supported by "computer chromatography software". But who can pay the costs of such a dedication?

"Single dedication" is not a MUST for computer chromatography, but lab net installation or just "big computers" do NOT automatically offer "computer chromatography".

1.1 The BIG Solution

The "big solution" also has big drawbacks—including the problems arising in use of excessively large software packages in big systems—whereas the small solution has so many benefits that the era of single dedicated computer use in chromatography can be said to have already begun.

As the validity of these arguments extends way beyond the confines of the chromatography laboratory, it is appropriate to repeat the author's remarks in the first issue of CAL—the International Journal for "Computer Applications in the Laboratory" when it was launched in 1983:

A revolution has just occurred in the laboratory!

Although computers themselves have become commonplace in the laboratory environment during the past two decades, the recent advent of personal computers has meant a gigantic leap forward with regard to speed, quality, accuracy or "cleverness" of everyday problem solutions. The single dedicated microcomputer

- provides ready access to incredibly large memories;
- enjoys almost unlimited software support and flexibility;
- permits use of powerful mathematical models to handle and solve laboratory data and decision problems.

Together these features open up a whole new field of activity, *viz.* dedicated problem solution under direct, critical user control.

Although large central computers might offer better programming languages, better programs, 32 bit precision instead of the 16 or 9 bit mathematical precision provided by many micros, faster access to gigabytes, much quicker job solutions than obtainable with micros, it is the direct access to computer aid which accounts for the present breakthrough.

This surely has something to do with human thought processes. We loose up to 75% of new information within 10 seconds of its acquisition. If the solution of a certain problem is going to take half a day (or just half an hour), then many of us "go routine" and quit (or must quit) deeper thought. If we are unable to optimize information immediately, or process it immediately upon production then we again quit (or must quit). If additional correlated information is not available at the instant we wish to consult it for more critical evaluation, or just for correlation purposes, then we simply forget it (or must forget it).

1.2 Brainware

Now, however, for the first time in the history of physical chemical analysis, the user is in a position to develop his own solutions—*i.e.* his own specific software—which is every bit as qualified as the instrument or laboratory tool he is using. We can call this new specific user-made or user-optimized or user-oriented or just problem-specified "product" brainware. Microcomputers have now created a situation in which "brainware" is instantly available in practically any laboratory.

1.3 Data "Risk"

Some colleagues are perfectly prepared to accept lab data accuracy and repeatability of about 1%, and that for just a single or at best a duplicate determination (in most cases on a single sample). However, all statistical tests reveal shockingly high levels of data risks inherent in single to duplicate measurements performed on a single sample by methods of around 1% repeatability. A few simple calculations will soon make it clear that our data cannot be too repeatable and accurate when viewed against the risk of making a false decision, with potentially drastic consequences, and against the cost of obtaining information using today's complex and expensive laboratory tools.

The workload in one of today's control laboratories or the complexity of work in an up-to-date research laboratory now has to be viewed in relation to the possible additional workload incurred on introduction of personal computers, or single dedicated systems, or independent but on-line computers.

The possible additional costs of such improvements have to be compared with the costs plus limitations of more highly integrated computer solutions.

1.4 Dedication versus Integration

Integration is the "only way" many "big" computer producers claim. But dedication (in the laboratory!) is so very much more powerful than integration—this is the lesson we have learned in the past ten to twenty lab computer years—that we are heading for disaster when we use bad models of "overall risk to overall cost" relations and when miscalculating the manpower necessary to change the laboratory structure. A new way of thinking seems to be necessary, after we have learned our lesson of creativity based upon satisfaction versus one-knob routine without any real relation to results and numerical values.

In addition, we are faced with "higher order" problems. Problems not just related to the laboratory, but to mentality, group interests, social structures, job qualification, and the acceptance of continuing education.

Let us just think of problems simply caused by lack of information: We sometimes appear to be divided into two camps: on the one side there is a large group of colleagues, fully aware of what big computers have had to offer for their laboratory work for many years and on the other side we find a small group of colleagues knowing precisely what single dedicated, small and powerful, programmed personal computers can offer in addition.

In fact, "in addition" is the operative expression. Dedication is not just a technicality, and therefore defies simplicistic technical analysis. Big computer users are well acquainted with the costs and limitations (of their system). This knowledge, although strictly limited to the "big" computer, often makes it impossible for this group to see, believe, or to seriously consider the utility of dedicated (and "small") computers in practice, although they might have heard of the many "aha" or "unbelievable" remarks uttered by the early programmers of the "micros". (In fact: none too few of them were also the programmers of the big computers as well!—who else could have started programming the micros?)

Well, lack of information due to lack of experiments and experience is one reason for a majority of colleagues to feel safe and sound in the hands of those who (only) know big computer possibilities. They cannot even overcome their inherent fear of wasting their time by touching a personal computer; such an attitude can consolidate prejudice.

Neither group of skeptics, the majority of today, which will become the minority of tomorrow, and the computer users of the 1960's and 1970's without access to the micros of the 1980's, can believe in dramatic improvements to their laboratory life. Seldom have groups of intelligent people been more polarized in technical matters than at this time of transition between integration and dedication. Quite apart from technical considerations, it is the human factor which plays a key role in this transitionary period. Dedication cannot be discussed merely in terms of 64k RAM memory, or Motorola 6502 chip architecture, (or its follower now

68000, a most successful 16 bit chip) or a Zilog Z80 (or its 16 bit follower) and the mnemonics table, etc. It is not enough to compare single dedication with the result of a lab data net concept, and show by cost analysis that the lab net system is superior. All this is lacking in "brainpower".

Some companies have closed their doors to anything containing an 8 bit chip as central microprocessor. They will get into trouble. The 8 bit chips will continue to do the job (see Apple model IIe starting in 1983). It is not the 8 bit that matters. The managers responsible for such policy are thinking only in hardware—they still have not got the story.

This laboratory revolution is one of attitude not hardware. Nevertheless, hardware is needed, and if the 16 bit based system, or the 8/16/32 bit tandem system still allows single dedication from the standpoint of costs, and if intelligent software is available, why not use it?

In the laboratory environment we shall soon be needing megabytes at each individual workplace. But we shall need to have this megabyte capacity instantly accessible round the clock at the point of application, in order to cope with all the raw data. And we shall need it at a reasonable price.

Independent, additional, and personal computerization down to the single instrument is dismissed as nonsense by a few instrument manufacturers. They forecast that this development will soon jar to a halt. Claiming that each new instrument is already computerized, they cannot believe in any change in the "one-knob mentality" cultivated so successfully in recent years. They cannot believe in any better software than that produced by their own software engineering department, where everything is directed towards well designed subprogram blocks easily assembled to satisfy any desire a customer may have. They think in terms of computers which offer optimized software packages to their customers, but in so doing they overlook, underestimate, or just deny the existence of a free person's "brainware".

I completely disagree with their forecast—simply because I am an analyst. My proposal is: make your instrument as sophisticated as you will, but be sure to add a good interface for easy access by 8/16/(32) bit personal computers.

Fully aware of the vast amount of additional work that has to be done by anybody entering the realm of intellectual freedom by opting for personal computer use at his workplace, I know that it is the guarantee of daily satisfaction which will help him win through. One-knob instrument lab routine is most frustrating, especially when we are confronted with those false results which looked so good before we discovered that excellent repeatability has nothing to do with accuracy.

And then there is the concept of "data cross correlation" This is something that emerged from the domain of the laboratory, not from the mind of a professional software writer. Data cross correlation is a sample related, or strictly problem related, approach to data based upon the laws of analytical chemistry, the laws of electricity, the laws governing the reactions of functional groups, or just the law of conservation of mass, *etc.,* known intimately to the practical laboratory worker.

Although there is a hard road ahead of everybody who wants to become truly independent or much better qualified in his laboratory work by using his brainware, there will be no regrets at the end of the day.

Computer application in the laboratory has much to offer all those who have taken, or are contemplating this step. So here is our invitation—let us tread this road together. The era of the single-knob instrument has gone for good; now's the time to start working with the 52+ knob instrument in the laboratory. You will master it.

2 Computer Chromatography?

Definitions are given of the terms "chromatography" "chromatographic data", "raw data and resolution of analog-to-digital signal conversion". Models for "raw data interpolation" are discussed as well as arguments for dedicated computer use in the chromatography laboratory. Qualitative and quantitative aspects of "computer chromatography" are discussed.

What is computer chromatography?

Computer chromatography optimizes the physical and chemical status of a chromatographic instrument. It optimizes stationary and mobile phase selection. It improves quantitative results by detection optimization, signal linearization, and noise treatment. It corrects retention time drifts, corrects for temperature and pressure changes, and makes for safe identification. It combines data from separation systems and detector channels in parallel operation.

In HPLC especially it helps to find the best mobile phase mixture and in HPTLC especially it removes the structure signal from the substance signal. Besides improving repeatability it ensures "accuracy". Analysis becomes faster and less expensive.

Concentrate:

Computer chromatography is far more than just computerized chromatography, which was itself a big step in the direction of more accurate and more repeatable results in chromatographic analysis. In the late 1960's computerization of GC and GC/MS was pioneered by E. Ziegler, G. Schomburg, and D. Henneberg at the Muelheim "Institut fuer Kohlenforschung" in West Germany. Computer chromatography improves even further gas chromatography (especially in smaller laboratories), helps very much in high pressure liquid chromatography, and is an essential prerequisite for high performance thin-layer chromatography. It has less to do with hardware, it is more brainware, which itself is much more than just software.

One can use computers in chromatography, one can computerize chromatography, but this is by no means equivalent to "computer chromatography."

In order to clarify what the term "computer chromatography" really implies, we shall need some definitions. However, since definitions often prove to be rather "dry", let us turn instead to some examples taken from real-life practical chromatography.

2.1 Definition of Computer Chromatography

Computer chromatography optimizes the physical and chemical status of a chromatographic instrument.

Optimization is a must if the BEST is to be made of a given chromatograph and the sampling technique, sample inlet, chosen column, and detector. However, even before choosing the column—stationary phase and mobile phase—intelligent computer programs might help to optimize the stationary AND the mobile phase selection. This volume does NOT discuss optimized phase selection; this topic will be considered in subsequent volumes.

Computer chromatography allows optimization of quantitative results, not only by mere integration of the chromatogram. Before integration one must OPTIMIZE the detector(s), the separation, the baseline, the signal (noise, drift). One must LINEARIZE the detector signal since there is no such thing as a chromatography detector showing a linear response over the full working range.

Computer chromatography helps to correct identification, to correct retention time drifts or changes, to correct for temperature changes, to intelligently combine data found on more than one column, to check for the "unexpected" side effect.

Computer chromatography in HPLC helps to solve the hardest problem most users have to face: which solvent mixture?

In TLC, computer chromatography corrects the worst problem of quantitative TLC: the layers used have a "structure" (measurable as signal peaks and spikes caused by the layer particles), which fools most of the existing electronic integrators. As a result, TL chromatograms WITH structure can be quantitized at best to within around one % rel. standard deviations, although structureless TL chromatograms can be quantitized to better than 0.05% rel. standard deviation. Some chromatograms cannot be quantitized under any circumstances using classical techniques, but furnish acceptable data when treated by subtraction of structure from the raw chromatogram. Detectability limits are found to be a full order of magnitude lower when using computer chromatography in instrumental TLC–HPTLC.

Computer chromatography allows facile improvement of any retention time data by at least one order of magnitude, thus remarkably improving qualitative and quantitative data, as the latter depend on correct identification.

There are so many further positive aspects that it would vastly exceed the space available here to summarize them all. However, we can begin to see that computer chromatography is really much more than just using computers in chromatography: it is not just a technique, but a whole concept.

If you feel you need some further definitions with respect to chromatography as such, then the following compilation has been prepared for you.

Otherwise GOTO >> Time Axis !

2.2 Further Definitions

- CHROMATOGRAPHY

Chromatography is a most versatile instrumental technique of substance mixture analysis by separation and quantitation of the separated parts of a sample.

Although chromatography can also be used for preparative separations, this volume does NOT treat any of the preparative aspects of chromatography.

- CHROMATOGRAPHY DATA

Chromatography is extremely data intensive. A chromatogram, however, consists only of peaks having a certain peak shape, showing a certain peak width "b0.5" at half height and a non-adjusted (raw) retention time "tms" corresponding to the tip of each peak, and a peak area "under" each peak, which is enclosed by a baseline, peak shape lines, and—in the case of incomplete separation by separation lines. This is easy to state but hard to measure—OR NOT MEASURABLE. Baseline and separation lines are INVISIBLE. It is the art of correct quantitation to find the baseline and separation lines as accurately as the peak shape lines are given. Chromatography signals are measurable as analog signal changes over time.

Analog signals can be digitized. The accuracy with which an analog signal is converted into a digital signal depends on the "resolution" of the "analog-to-digital converter", the A/D transforming system which is an essential component of any computer chromatograph.

2.3 Raw Data and the Resolution of A/D Conversion

The lowest acceptable resolution for chromatography is 12 bit resolution, which means that the smallest digitized signal represents one bit and the largest 2 to the power of 12, which is 4096. However, we are not dealing only with the peak height value, but also with the TIME corresponding to each raw data point. The time axis is of paramount importance in chromatography for it is there that the qualitative information resides. A 10 millisecond absolute accuracy of the time component of a raw data point would seem to satisfy all needs in normal speed elution chromatography. How time data can be stored in the storage of raw data point sets will be discussed in Chapter 3 "Time Axis".

The peak signal can be stored as a continuous line written by recorders or may be quantitized into discrete "raw data points".

● MATHEMATICAL MODELS for DATA INTERPOLATION

If as given in Figure (1) about 8 "raw data" points represent a peak, mathematical models for raw data interpolation are needed to reconstruct the true peak shape when replotting digitized data back into an analog figure.

The result is seen in Figure (2). Here "cubic spline techniques" have been used, to obtain usable data. We will discuss this important procedure later on in detail.

These methods allow reduction of raw data storage without loss of accuracy especially in wide range isocratic LC or isothermal GC. Here very narrow peaks and very broad peaks are combined in one data set.

In order to obtain accurate peak maximum data with respect to peak height and corresponding time, higher

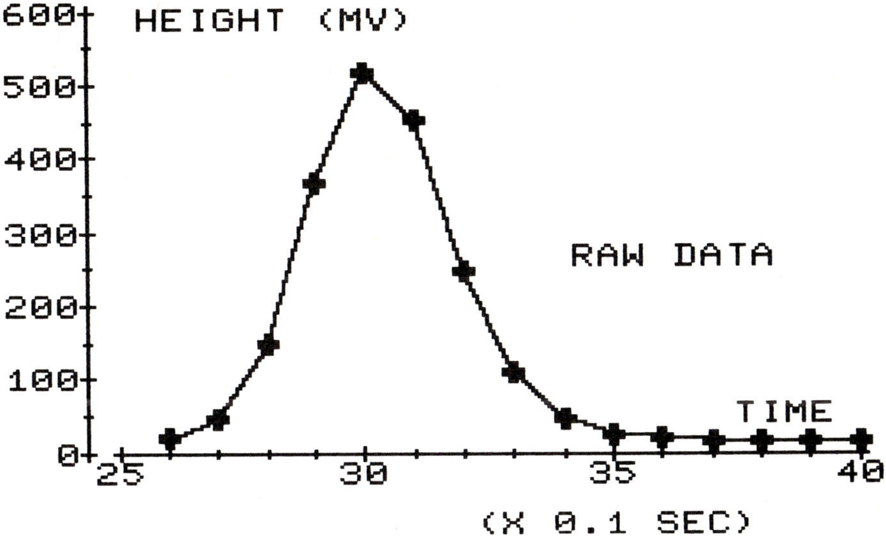

Figure (1)
Raw data: X-axis = time in 0.1 second units. Y-axis = A/D conversion signal in units. Software used for the Figure: SCIPLOT by P.K.Warme—see APPENDIX SOFTWARE.

order polynomial raw data interpolation is necessary. This allows calculation of the precise retention time value of any peak which has been digitized.

If more than 20 raw data points per half height peak width (b 0.5) arc mcaourcd for caoh cinglo coparatod peak than raw data storage tends to become "overdone". Chromatographic techniques are known which need 100 and more raw data points per second. Most real-life chromatography can be treated well by storing .1 to 10 data points per second.

NOTE: How many data points per second are necessary will depend on the smallest peak width at half height starting at that part of the chromatogram where qualitative and/or quantitative evaluation is desired.

But it is not always the density of raw data points per second—the AD sampling rate—which is of interest. In some cases of computer chromatography it is the density of digitization in "points per mm". But here again dynamic consideration play a role. Distances in "Layer

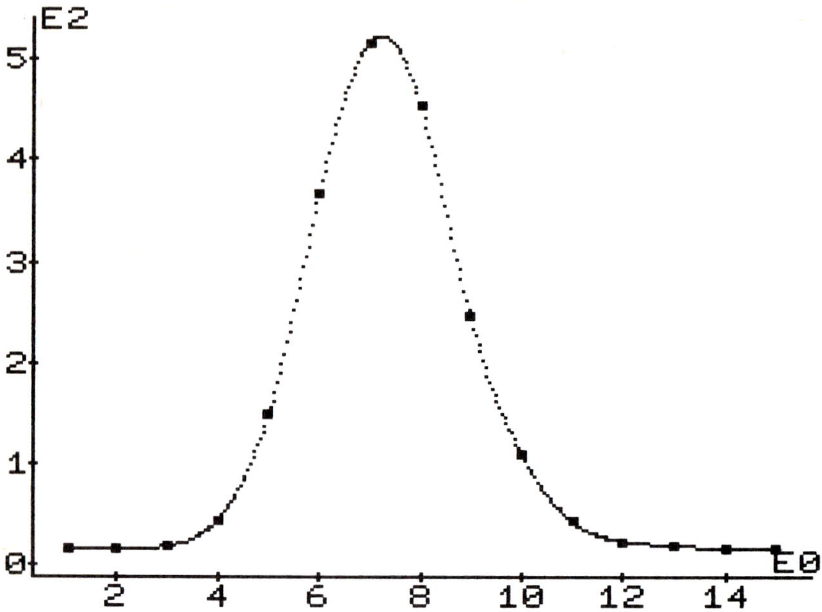

Figure (2)
The raw data points in Fig. (1) have been connected by a "cubic spline" interpolation line. Software used: CURFIT by P.K.Warme—see APPENDIX SOFTWARE

Chromatography" can be translated into movement. From this speed and final digitization rates are to be considered:

Instrumental thin-layer chromatography, for instance, can furnish the whole qualitative and quantitative information forthcoming from about 20 to 70 samples—each consisting of 5 to 20 chemical species within 3 to 10 minutes, thus producing 40,000 to 140,000 data "points" of at least 12 bit resolution within this short time.

The chromatographic data based upon raw data sets are not easily read but are highly amenable to treatment by such techniques as interpolation, correlation, mathematical smoothing, integration, or graphical display, provided that cleverly programmed software is available for running in computers which can ensure fast processing of large amounts of data. Large amounts of data, high speed of data production, intelligent data reduction call for direct computer support.

As chromatographic instruments became qualified tools in the mid-1960's just when computers became powerful enough to handle huge data sets, computerized chromatography is as old as the technical use of computers. It was the sheer volume and variety of the analytical problems arising throughout the whole wide field of chromatography applications—from gas, through high performance liquid, to instrumental thin-layer chromatography—that shifted computerized chromatography towards "computer chromatography".

If you NEED MORE ARGUMENTS about why computers should be used in chromatography, here they are:

Some analysts have claimed that using a computer in primitive thin-layer chromatography must indicate that something is wrong: either with the TLC user or with TLC. There is an element of truth in this claim: we would never have used computers in TLC; the only sensible approach is to make TLC instrumentally suitable for use in connection with computer—and suddenly we end up with a most versatile, cost effective, fast, and powerful analytical technique.

Well, this is easy to state but perhaps not too easy to understand, so you might really want some more arguments.

If NOT, GOTO >> Time Axis.

2.4 Why Use a Computer in Gas Chromatography? Or in HPLC or HPTLC ?

Like HPLC, GC has an "open time axis". This means that qualitative analysis of chromatographic information is much more complicated than for any other instrumental analytical technique, such as spectroscopy, titration, polarography, etc.

Although a chromatogram consists only of a limited number of "peaks" above a baseline, and although there are only two aspects of chromatography data analysis, *viz.* qualitative and quantitative evaluation, the final task is complicated and prone to systematic errors.

The most elegant gas- and liquid chromatography instruments are now available, or feasible, which are already optimized. They have perfect temperature control down to or even below the 0.1 °C level. They have pressure or flow control devices based upon fluidex principles, keeping the mobile phase flow constant within a 0.1% range or even better. Temperature and flow can be programmed in any mode and are repeatable to much better than 0.1% rel. Detectors are linearized and work at many selectable channels of different mode or wavelength, simultaneously or serially. More than one separation systems runs under optimized conditions and the optimization has been correlated with the type and composition of the sample to be analyzed or at least this is all in preparation.

Time data of chromatography signals are scanned at an accuracy level of below 10 milliseconds per raw data point. All raw data are stored electronically and interpolated by the practical optimum of the available mathematical models.

Systematic cross correlation of chromatography data is used to check for plausibility in any detailed result; autostatistics is used to check for the weight of any alarming unexpected analytical result and to condense other results to the minimum necessary without loss of information, thus making the analyst's life easier.

A most intelligent data reduction system analyzes final results. It checks for the possibility of proposing more selective or more accurate and precise improved analytical procedures.

Self error check, self calibration, self optimization, even self repair where possible is done automatically (or at least is planned to be introduced) and daily thus reducing costs otherwise incurred by bad results ...At least this could be done with the most advanced instruments in gas and liquid chromatography.

However, there are many old instruments in laboratories, which cannot be exchanged for the finest new tools, or where the daily load of routine analyses simply has not allowed improvements, modernization, better tools, or better techniques.

Since this is the true situation prevailing in the vast majority of cases, the problems of bad analytical techniques, or the effects of weak instrument should be discussed as follows:

2.5 Qualitative Aspects

The quality of the substance represented by a peak is given by the time corresponding to the peak maximum.

The position of this maximum depends on:

- the mobile phase flow

- the temperature or temperature DISTRIBUTION along the column

- the inlet pressure, outlet pressure, and pressure GRADIENT along the column

- the length of the column

- the particle size, particle distribution, pore distribution, and mobile phase distribution along the packing in a column or—in the case of capillary chromatography—along the capillary tube.

- Even slight changes of the weather conditions in a laboratory will affect the time data of chromatography peaks.

- Peak maxima and thus the time data are influenced by the peak shape. The peak shape varies with the conditions of sampling (pressure effects, temperature, electric charge effects) during evaporation of injected non-gaseous samples.

- The positions of peak maxima are influenced by neighboring peaks.

- The peaks may represent MIXTURES of differing substances

- The range of quantities of mixed or superimposed peaks may vary from ppb to whole percentages.

The list of factors influencing the time positions and shapes of peaks is almost endless, but a correct and precise value for the peak maximum position is the ONLY means of ensuring qualitative analytical information.

Of course: it is well known that the majority of chromatographers do not need qualitative data or believe that they do not need them. However, it is better to use qualitative data. They are THE ONLY qualified "address" to correlate a correction factor with a peak number or name. Computers need addresses of this type! The analyst simply looks at a chromatogram and then puts together peak name and correction factor—not so the machine. Qualitative accuracy is therefore the basis of quantitation!

2.6 Quantitative Aspects

- Quantitative data are deduced from peak area measurements. One could simply regard area integration as a procedure for quantitative analysis. However, the peak area cannot easily be measured as a whole series of factors introduce complications:

Basically, NO information is actually available about the true position of a baseline under the peak, as elution techniques like GC and HPLC are single consecutive methods allowing no simultaneous measurements of the "zero chromatography condition" prior to the real chromatogram. The analyst can only hope that the baseline is an interpolatable line. He can just trust that he will find sufficiently large sections of a chromatogram which are "empty", *i.e.* which are free from substances. However, there is little justification for believing in empty spaces within a real-life chromatogram.

In many real analytical situations it is illegal to use the position of a baseline prior to a chromatographic run—*i.e.* the zero signal condition—as the true baseline of the actual analytical run. Is the detector zero signal position independent of changes of the carrier gas flow? Hopefully NOT.

During the real analytical run the mobile phase flow through a column very probably will differ from the flow which exists with pure mobile phase—*i.e.* without any sample.

- Many peaks overlap. Where is the true "separation line"?

- The detector calibration line will be non-linear. There is no truly linear detector, giving a linear response over the entire range used in practical analysis. Thus integrators integrate signals which may differ dramatically in their signal-to-mass ratios.

- Each peak encompasses the lowest trace level of concentration to the maximum concentration.

- In each case of quantitation of substance mixtures under a peak, the specific detector function signal expressed in grams/second or grams per volume of substance/mobile phase is influenced by the fact that a mixture of substances is present.

Let us just consider the case of traces of hetero atoms in a special chemical bond superimposed upon a main substance signal when a flame ionization detector or an electron capture detector is used in GC or the case of special trace impurity conditions in the mobile phase during a liquid chromatography UV-detection run: these are specific effects, affecting the specific calibration conditions. If they are unknown, they cannot be simulated.

These and many other effects not yet mentioned are the reason for the present, most disappointing situation existing in analytical chromatography today, in the 30th year of its practical application. Only the computer can help.

GC and HPLC produce quantitative analytical results which can be reproduced to within hundredths to a few tenths of one percent relative standard deviation, but are subject to systematic errors which are all too often ten to one hundred times larger than the repeatability.

Here are some of the reasons for this really shocking situation, shocking in as much as chromatography is today's most widely used analytical method and is employed to obtain data which represent the very basis for legal proceedings, mass production decisions, environmental activities, *etc.*:

- The temperature in GC instruments is measured by primitive techniques, often at the worst local position, by instruments which cannot be read as precisely as necessary.

- Temperature gradients in GC instruments are subject to change, depending upon the temperature level itself, and often suffer irreversible changes after a high temperature shock.

- In HPLC most of the analytical runs are done without any temperature control. Do not believe that temperature cannot influence reversed phase LC! Don't believe it even when you repeatedly read about this nonsense. Just check it with your system and your sample. And don't be shocked by any sudden disappearance or by an entirely new peak never seen before in your "familiar" sample!

- Mobile phase flow in GC is possibly controlled—if really under control—at the worst possible point within the separation system.

- Mobile phase flow in HPLC would be kept constant, if everything else were to remain constant, by the mobile phase pump characteristics. However, 80% of all HPLC runs are performed under changing mobile phase conditions and here are some "side effects" widely unknown.

● During a chromatographic run, the whole chromatographic system changes in as many steps as there are peaks separated, according to the chemical and physical properties.

● Detectors are basically not linearized, at least not for YOUR specific compounds under your conditions of the mobile phase true composition.

● Integrators are basically not fed with linearized signals but non-linear signal integration is illegal if not quantitatively under control.

● The time characteristics of most of even the latest integrators are far from being correlated with the time characteristics of the latest separation systems, thus in fact "killing accuracy and precision" of qualitative chromatographic data.

These are just a few of the "bad highlights" characterizing qualitative and quantitative gas- and column liquid chromatography today, now that dedicated computers have become available at affordable prices to help and to dramatically improve the situation.

In thin-layer chromatography the overall conditions for precise and accurate analysis are even worse than in GC and HPLC.

In fact, many analysts start forgetting TLC as a valid analytical tool, as due to the use of electronic integrators the accuracy and repeatability of quantitative TLC data was lowered by one to two orders of magnitude especially in the trace analysis range.

When quantitizing a TLC plate under a photometer scanner one finds an excellently reproducible, so-called plate layer structure signal as the baseline under a chromatogram. In contrast to GC or HPLC, where baselines cannot be measured prior to the chromatographic run under the conditions of the chromatography step, TLC offers the possibility of baseline "analysis" prior to the final chromatographic use. Qualified scanners show a very small noise signal compared with the structured signal of the empty layer, a signal, which unfortunately was called "noise" and was rejected or suppressed by strong signal filtering. If one feeds TLC scanner signals into electronic integrators, which have been programmed to handle elution chromatography signals, the integrators are fooled by the structure signal and dramatically lower their peak detection limits and enlarge quite dramatically their noise rejection factors.

This is the reason, why a TLC data set, which can be reproduced down to all details at repeatability levels of below 0.04%, can be integrated in electronic integrators only at a level of around 1...2% repeatability at best.

Therefore: computer chromatography.

3 Time Axis in Gas Chromatography

GC is "signal over time" only, or in other words: there is at first just a simple chromatogram, nothing else. GC is a serial single shot-one sample analysis, if done simply. There is no further information forthcoming, if GC is considered in isolation, i.e. on looking at ONE chromatogram, ONE column, ONE detector only, as such. At a first glance, the result of a single analysis in ONE run would appear to consist of the non-adjusted retention times of the peak maxima as the only indication of a substance "name" and the peak area (signal times time integral data) of 1 to 4000 peaks. It seems that about one percent of the quantitative data of this small part of ONE chromatogram is all we need.

A closer look shows, however, that RELATIVE, or (even better) relative/relative time data can considerably increase the GC information content.

The really dramatic breakthrough in GC information is effected by something, which is possible ONLY with computer chromatography:

1) The fact, that the WHOLE chromatogram data can be stored as a full set of raw data. Any data treatment, any procedure to extract information from a chromatogram, or to compare it with others, or to adjust it, to correct for pressure effects (= weather!), for temperature drifts, for sampling side effects, for...etc., can be done without affecting the original data. Integration can be repeated under improved conditions, at much higher speed than during the real analysis...

2) An improvement of time data precision by one full order of magnitude on using powerful and sufficiently accurate fifth order polynomial time bit interpolation techniques, which permits highly precise, qualitative routine GC.

3) Automatic peak width and peak shape analysis of high level quality by combining interpolation techniques in order to use the peak shape/peak width/time correlation data for physical separation improvements, for selectivity adjustment (= overall optimization of the GC separation).

4) Intelligent data comparison when using more than one sample, more than one inlet/outlet of one column, or more than one column (not yet multicolumn) or more than one detector, or when SIMULTANEOUS analysis is done. Even when doing the SAME analysis of ONE sample with ONE column and detector but at MORE than ONE temperature, extremely useful information can very often be found, IF the DATA are EVALUATED to the real LIMIT of evaluation. This is necessary in order to really identify separated peaks, or to justify quantitation, or to provide peak purity data.

5) System analysis of the GC system for correct column qualification or column selection, injection and instrument optimization for maximum yield of information in minimum possible time or at minimum possible cost.

There are OTHER aspects, such as detector optimization, or signal linearization, or optimized chromatogram display, or intelligent GC data reduction, or database optimization, or instrument improvements, or statistical experiment planning or elegant AUTO-REPORTING including auto-statistics, which are likewise possible ONLY by computer chromatography.

In the following sections we will discuss TIME data-dependent factors only and will treat further aspects in a subsequent volume. The KEY RESULT of this section will be: Time data in chromatography cannot be measured (or calculated) precisely enough. A 10 millisecond precision IS A MUST when using multilevel evaluation. Computer use is a must when evaluating GC data to the full. Computer chromatography is the ONLY way to reduce fully evaluated, intelligent GC data to human readability for just one purpose:

To obtain qualified results.

Concentrate

This section deals with:

- definition of time data in chromatography;
- which "raw time data" to measure and how to do it;

- **how to store "raw time data";**
- **how to use computer chromatography for best interpolation of raw time data to get accurate retention time values;**
- **why time data must be improved to a precision level of 10 milliseconds and the benefits of this (easily accomplished) improvement;**
- **how to get peak width data of high precision and accuracy.**

Retention time "tms" (s) is the raw (non-adjusted) time elapsing between the instant of injection and the elution of a peak maximum.

Adjusted retention time "ts" (s) is a retention time value, corrected by subtraction of a so-called dead time value tm. ts data are obtained by calculation only.

Dead time "tm" (s) (in contrast to GC unit definitions—for many good reasons) is an important practical time value. However, it can only be obtained by calculation.

Computer chromatography offers a useful algorithm to calculate dead time data, which are THE measure of mobile phase flow in elution chromatography.

Peak width b0.5 (s) at half height (of a peak) is a value important enough to be calculated. You will miss the "peak width, w" at the basis of a peak. We have serious reasons NOT to use "w" values. In practice they are useless. Their "theoretical basis"—which says: "w" equals four times sigma—is simply wrong, as there is no Gaussian function.

Computer chromatography offers now accurate b0.5 data; however, it is the TRUE value of b0.5 which counts. Some commercially available data handling systems in chromatography—including the most recent ones—calculate a peak width value based upon assumption of a certain peak shape model (Gaussian).
These "indirectly" calculated peak width data are useless, especially at the beginning of a chromatogram. These values deviate remarkably from the true peak width at half height data especially at the early part of the chromatogram.

b0.5 values are important for the automatic optimization of the physical part of chromatography separation, but also offer a lot more besides.

Indirect time data are correlated with peak shape, baseline position, separation lines and their positions. They can be calculated by certain computer chromatography algorithms, but belong to second class time data. Nevertheless, they are important enough to warrant attention and will be discussed later.

ALL TIME data in elution chromatography are given in seconds with a precision of ±0.01 second. Throughout the whole series of volumes on computer chromatography, we will avoid the use of time related data like mm, cm, inch, *etc.* or minutes when evaluating chromatograms in contrast to some integrator data which can be presented in minutes, 0.1 and 0.01 minutes.

Again it is stated that time data in chromatography MUST HAVE a precision level of below 0.1 s, preferably at the 10 ms level. This is just as true for normal everyday ROUTINE CHROMATOGRAPHY.

Many readers will disagree with this statement. They will have ample experimental evidence of their own, showing—at a first glance—that there is NO need for or NO instrumental qualification for a 0.01 second precision in elution chromatography. They will state that one CANNOT measure peak maximum data to better than 0.1 to 1 second precision and will say that they do not or cannot use such a precision.

They will claim that it is impossible in routine practice to inject the sample with an accuracy within one tenth of a second and they will state, that no pressure regulator or no thermostat will permit work at 0.1 second precision. This is all very true...

3.1 Data Cross Correlation: Time Accuracy

The 0.01 second accuracy is necessary not only in case of fast chromatography runs. It is also needed for one of the principal benefits of computer chromatography: DATA CROSS CORRELATION! We shall demonstrate, with the aid of examples, the value of data cross correlation not only for identification purposes in chromatography, but especially for excluding false experimental values which so often falsify chromatography results.

With respect to FAST chromatography we have already entered this area:

HPLC with 3 micron columns and capillary GC with 0.2 mm fused silica capillaries or low stationary phase impregnated packed columns in GC produce fast chromatograms corresponding to only 300 to 600 seconds analysis time.

One should realize that it is the ADJUSTED retention time that counts, not the raw (non-adjusted) retention time, in HPLC and GC. The first eluted peaks may easily have adjusted retention time data of a few seconds only. And it is THIS part of the chromatogram which needs a 0.01 second precision to avoid loss of analytical information.

In instrumental HPTLC a "0.01 second precision" is totally inadequate! Here we need even faster timing. No problems with computer chromatography: An Apple II plus the proper A/D converter can handle 15,000 digital data at 12 bit resolution per second! But here we are still discussing only 10 measurements per second, not 10,000...

Computer chromatography techniques will show that:

- we can achieve this ten millisecond precision in practice;
- we can overcome the problems of non-accurate start conditions with respect to time;
- this precision is of real practical value.

The following statement is valid over the full range:

3.2 Time Data Cannot be Precise Enough in GC and HPLC

SOME MORE ARGUMENTS—if you do not wish to read more now,
GOTO >> Benefits

In HPTLC time data are mostly position data. So we must include "position data precision" for HPTLC in the key statement given above.

Some analysts consider a biosphere sample repeatability at the 4-10 % level as acceptable. This, however, reflects a deep misunderstanding of what the statistics of analysis have to say. The single biological unit IS itself highly precise at a given moment. Here too, analytical data must be accurate at the 0.1% level of repeatability.

As computer chromatography can offer this quality-of-analysis level we cannot afford to neglect this opportunity—because the following is also true:

Analytical data cannot be too precise and accurate, as the statistics of real life analysis—we use only a very small number of samples, a very small number of repeated analysis of these few samples—show a shockingly wide range of uncertainty of the final result.

Anybody who persists in rejecting this statement should spend a minute studying Figure (3). It shows the correlation of

- method precision

with

- number of repeated measurements

and

- risk of the final result.

Figure (3) is the accepted statistical t-test in analysis. It clearly shows that the best possible method quality given as the standard deviation is necessary to get acceptable real life data when using just only 2...4 repeated measurements, as is done in practice. Figure

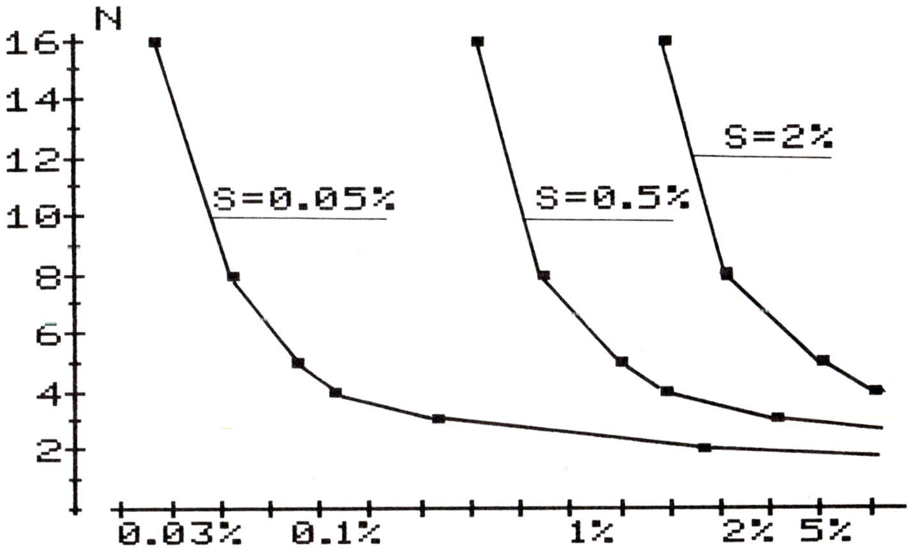

Figure (3)
Statistical t test. Y axis: number of repeated measurements. X axis: risk of resulting error in %. The three lines represent three quality levels of methods, given in the standard deviation of repeated data.

(3) also helps us to understand why the analyst should use the best possible tool giving the highest possible repeatability because of this statement:

Most of us use only TWO measurements. One original, one repeated (not to mention "analysts" "measuring" simply once and with one single sample from just one source!!). Figure (3) tells us about the risk of duplicate measurement when using a "2% repeatability level technique"!

However, repeatability is only one side of the story. There is also accuracy, which counts all too often in practice. CROSS CORRELATION of data is THE method for reducing the danger of getting "excellently repeatable but totally wrong" results.

Cross correlation of data is so data intensive in chromatography that the technique is largely unknown, as it could not hitherto be used in practice. However, computer chromatography "can do it". It allows—and therefore uses—"chromatography data cross correlation". To illustrate what this means in practice, let us consider just one example from the realm of gas chromatography:

There is a linear relationship between molecular weight of a substance and the logarithm of the true adjusted retention time within a homologous series of non-polar compounds.

Although—when looking at fine details—the line has a "zig-zag" shape in the low molecular weight range, it becomes less zig-zag from substance to substance; however, the mean of this "zig-zag" is linear over the whole range of molecular weight covered by gas chromatography. The "zig-zag effect" is a manifestation of the odd-even molecular structure in consecutive homologous series.

The practical chromatographer might feel that zig-zag lines, homologous series, log data, *etc., etc.*, have nothing to do with his real-life sample and his problem of getting correct quantitative data here and now. But there is not only "cross correlation" of data in chromatography! Chromatography itself is cross correlated, from the surface of the bottle to the closing report. We can only invite him to come on in, instead of waiting outside. The facts will soon speak for themselves.

Again: Time data cannot be precise enough.

3.3 Benefits...

of computer chromatography with respect to time axis data treatment and factors to be taken into account prior to raw data measurements:

Time and signal height represent the only information immediately available. Therefore accurate time is important for qualitative AND quantitative GC analysis, is important for optimization of the separation, and is of utmost importance for ensuring a sufficiently high information yield per unit time. Qualified data acquired in the shortest possible time is a good return on investment in expensive equipment and will offset the ever-increasing cost of manpower necessary in chromatographic analysis.

GC data today are too often of the following quality: Extremely well reproducible, much too often systematically wrong, sometimes totally false (wrong identification, undetected substance overlapping). The costs of misinterpretation or misdecision based upon such GC data are high. This is stated by an analyst who has worked in and with GC, LC and TLC for 30 years in almost all fields of analysis, including industrial, environmental, pure research, forensic...

The statements are true for gas analysis or analysis at the very limit of technical feasibility of GC, from −180 to +450 °C and from 99% to one thousandth of a ppb concentration.

However, GC is or at least can be very powerful; yet it can also be done incredibly badly. The chromatogram does not tell us anything about the precision and accuracy of the results: but CROSS CORRELATION of data does in most cases.

Computer chromatography possesses an immensely powerful weapon against misinterpretation:

3.4 Cross Correlation of Data besides Time

And THIS is what makes it so necessary to get data at a 10 millisecond accuracy level. If you still have your doubts, then just wait and see that even 10 milliseconds accuracy will kill part of the precision of fast GC (as well as "3 micron particle" high pressure LC) analysis. And today's fast GC will become even faster tomorrow, because of the new possibilities computer chromatography has to offer.

A few further remarks with respect to high or low speed in chromatography are in order here:

- Excessively slow chromatography cannot be quantitized by electronic integrators, and excessively fast analysis cannot be registered by recorders, or integrated properly (accurately enough) by "normal" integrators.
- Excessively noisy chromatograms create a lot of trouble for recorders and integrators. What about clever digital filtering?
- Too poorly resolved peak groups cannot be handled by integrators. Why not handle "grouped GC" by a computer PLUS the on-line analyst, *i.e.* the computer's graphics display taken as an aid for decision by the analyst who knows what he wants?
- Better separation is our first priority. But what if we have already reached the limits?

We need GC signals which are sufficiently fast and yet sufficiently slow. We need them optimized. Ideally, all the peaks should be baseline separated and equally distributed along the time axis. The reader's first reaction might be to say "nonsense"? Consider the possibility of computer chromatography this way:

Starting with two columns—a non-polar and a polar one—both at equal temperatures and coupled in series, the computer starts a series of sample injections by commanding an autosampler. From injection to injection the temperature of the polar column is raised. The computer program detects the NUMBER of peaks and stores the retention time of each. If the total peak number drops from injection to injection, the program reverses the course of action: instead of increasing the temperature of the polar column it proceeds stepwise to lower temperature.

If the retention time of the last eluted peak becomes too large, the program starts increasing the temperature of the nonpolar column.

A hot polar and cold nonpolar column means a nonpolar series column separation system. A hot non-polar and a cold polar column in serial means a polar separation system. (It is the "in column" residence time in the real sense of retention time which counts [1].)

The computer program works towards MAXIMUM number of peaks and then, by second priority and fine tuning of two temperatures, towards the smallest standard deviation of the peak distance data (peak to peak). This smallest possible standard deviation value is reached when peaks are equally distributed.

Just as we said: the best possible situation would be to have all peaks baseline separated and equally distributed by time! We tried it in the way described above. It cost 14 days to do it manually. We plan to describe the algorithm used for this part of Computer Chromatography in the volume: "Optimization of Separation".

Chromatograms are frequently too fast or too slow, or the peaks are offscale or too small.

Consider this possibility: The GC signals, whether too slow or too fast, are stored on floppy disks and the stored chromatogram is then "replayed" at convenient speed into the equipment we have NOW. Let us discuss the equal distribution of peaks of real life samples later, in one of the subsequent volumes of "Computer Chromatography" and now concentrate on TIME only.

There now follow some final remarks about:
retention time
adjusted retention time
dead time

and what we can do with
cross correlation
in this field.

If you are familiar with the details GOTO >> "Experimental"
ELSE continue:

We need the

RETENTION TIME

i.e. the time distance from the moment of injection to the position of a peak signal maximum.

We need the time width of the peak signal at half height—*i.e.* the PEAK WIDTH at half height.

It is useful to consider the TOTAL of all signal-over-height data, the PEAK SHAPE data, and for quantitative evaluation it is necessary to find the

PEAK START and PEAK END TIME.

Finally, one can state that it is of value to store the total chromatogram, right down to the miniscule details over the whole time range.

Before discussing the measurement of time corresponding to peak maxima, we shall first consider the necessary precision (and accuracy) of time data measurement and calculation.

We have to compare
absolute time values with
relative time values
and
adjusted retention time with
non-adjusted retention time.

Although not widely used in the literature up to now I would invite the reader to accept the following nomenclature:

Let us consider chromatography as a process, where substances are separated due to DIFFERENCES in their "stationary phase residence" = RETENTION time. Let us simply assume that the time differences are caused exclusively by residence time in the stationary phase, while the mobile phase transfer time is more or less equal for all substances.

In fact, this is not strictly true if we examine the finest details; roughly, however, the following algorithm holds:

tms = total residence time of a substance in the chromatograph

ts = stationary phase residence time. ("s" is for "stationary phase")

tm = mobile phase transfer time ("m" is for "mobile phase").

Retention is caused by stationary phase residence time. (In fact: migration hindrance due to phase/phase transfer hindrance or due to migration hindrance caused by micropore structure is considered as belonging solely to the "stationary phase induced residence time".)

Now we can simply write
ts = tms − tm
and state:

Only THIS DIFFERENCE between two time components

is of importance for a separation. However, ONLY tms can really be measured.

Many chromatography users do not recognize the importance of this fact, because the majority of chromatographers are using temperature programmed techniques. And in TGC (temperature programmed GC) residence time data are changing continuously during a separation; they are highly dependent upon very many factors. In fact, the mobile phase transfer time tm, which from now on we will call

DEAD TIME

changes as the stationary phase residence time ts changes. Discussions of accuracy and repeatability of time measurements therefore depend greatly upon the GC technique used.

Let us initially assume that we are using ISOTHERMAL GC. And we use THIS technique for identification, data justification, and precise quantitation. Then the following is true:

There is no doubt that raw (non-adjusted) retention time date are useless in practice. Absolute retention time data have no meaning with respect to qualitative analysis. Non-adjusted, raw, retention time—we will use the symbol "tms" (seconds) for it—depends on temperature, carrier gas viscosity (which changes throughout the analysis, depending upon how much and which type of sample is in a column), inlet pressure changes, outlet pressure changes, which include the weather (!), and various other factors will all exert an influence on the non-adjusted retention time. However: all these influencing factors are easy to measure and so in principle even NON-ADJUSTED retention time data could become of value in computer chromatography.
In practice RELATIVE retention time data are of wider use but of no fundamental importance. Only relative ADJUSTED retention time data are comparable in case of differing conditions of temperature, column length, inlet pressure, stationary phase amount, weather, *etc., etc.* This is especially important in case of

a) isothermal GC
b) qualitative analysis.
c) quantitative analysis
d) data justification.

One will probably ask how adjusted relative time data can be in context with quantitative analysis?

Well, this has to do with certain effects already mentioned earlier, but as this is often unknown or considered as "academic" only it is justified to repeat here:

- detectors work non-linearly
- peak shape depends on retention time
- integrator data depend on peak shape
- data justification depends on correct identification
- and identification is possible ONLY on the basis of relative adjusted retention time data.

Many other important facts can likewise be expressed only in terms of adjusted or relative adjusted retention time values. The basis of relative adjusted retention time data "ts" (seconds) unfortunately is the dead time.
It is easy to say that this is the mobile phase residence time of any compound migrating through a column.
So we simply use the symbol "tm" (seconds) for the dead time:

tm = tms − ts (s)

(for any compound, at least at a first glance)
but
tm cannot be measured directly
tm cannot be calculated easily
tm data are prone to systematic errors.

For these and other reasons, many a gas chromatographer has been deluded into believing in "simple procedures" promising a short cut to the value of tm by using the retention time data of the "first eluted peak"; this amounts to saying that tm = tms and ts of the first eluted substance is therefore zero—which is never true.

In most capillary chromatography analyses, the adjusted retention time of methane is around 1...3 seconds. Compared to the adjusted retention time of 5...20 seconds for the next compounds, this is a remarkably large value. Remember: all comparisons and identifications are possible only on the basis of ADJUSTED (ts —) values, not the raw retention time tms.

The same is true, for example, of methylene chloride as "dead time" substance in GC with electron capture detectors. Identification procedures are doomed to failure on use of the tms value of such a substance instead of the real dead time value. We then have a perfect case of overkill: the user no longer sees any point in working with identification techniques in the case of short packed columns plus ECD. He rejects the best tool he has at his disposal to ensure correct results, because his integrators—with a precision limit of 1 second or at best 0.1 second—will have ruined his chances of performing qualitative chromatography. He is now firmly convinced that GC is entirely inadequate in this respect.

Computer chromatography can change his mind dramatically. Just a few experiments on the 10 millisecond level will suffice to persuade him of the importance and value of the dead time and to utilize it in his own work.

"tm" is the basis for qualitative work; moreover, it is THE measure of mobile phase flow through the chromatographic system.

Various procedures are known for the approximate measurement of tm, but none of them succeeds under routine analysis conditions as a consequence of the extra time and effort necessary.

3.5 Dead Time Cannot Be Measured Easily

Dead time data are not replaceable if we wish to exploit the full information content of GC data and utilize the enormous potential of data cross correlation. It is this technique of data cross correlation which is the basis of data justification, and above all the very foundation of a checking procedure for systematic analytical errors. It is mainly this latter point which makes CORRECT tm data indispensable. Furthermore: tm is THE measure of mobile phase flow and the flow data are in turn the basis of optimization of physical separation.

As most chromatographers never have time to optimize, AUTO-optimization is the only practicable way to utilize our tools to their limits in order to obtain the best results in the shortest time at the lowest possible cost levels.

It is THIS which makes tm data so important.

As the majority of chromatographers have never optimized their instruments, most of us cannot appreciate the benefits. (Who can really appreciate something he has never seen?) Here the reader is invited to persevere.

tm is obtained by calculation.

The algorithms available for its calculation demand the

utmost precision AND accuracy of the non-adjusted retention time data "tms" of a KNOWN mixture of substances. We must know the retention index data of our mixture, and for mathematical reasons we must use a mixture of AT LEAST four compounds.

So we need precise tms data of a real ISOCRATIC GC run. In order to obtain these data we shall require:

- a GC instrument, a clean test sample (see below for conditions)
- a chromatographer who knows how to inject a test sample properly and start a timer correctly
- (or an auto-injection device)
- before injection: waiting time to ensure constant temperature (available at night in automated runs, or available under isothermal routine conditions)
- a sample mixture, which has the solvent eluting as LAST PEAK (otherwise isocratic conditions—overall constant conditions of temperature, flow, constant polarity of mobile and stationary phase—are impossible.)
- at least 10 raw data points per peak width at half height—see, however, Figures (1 and 2) above which demonstrates the potential of "spline techniques" for reconstructing basic peak data such as peak TOP, peak WIDTH at half height, *etc.*)
- an AD converter of at least 12 bit resolution on line to the detector signal amplifier
- storage of at least twenty raw time data per peak
- a proper program which finds peaks, calculates the baseline, finds the peak top time data and stores just these data for later treatment (or handles the whole story in real time)
- a proper mathematical model for regression of the raw data points around the top of each test substance peak in order to improve the time precision by one order of magnitude—see Figures (2) to (5) and
- a "tm calculation" program

3.6 Hardware Conditions

The analog signal from GC instruments lies in the range of 10 microvolts to 10 volts, and is often limited to 2...5 volts maximum signal DC.

If we were to use a signal amplifier transforming 10 microvolts into 10 millivolts as a level ONE signal range and 10 millivolts into 10 volts as a level TWO signal range, and if this amplifier were software controllable, we would be in a position to handle two signal voltage ranges, and thus cover the whole range of output signals encountered in practical GC. Signals extending over a range of six orders of magnitude could then be sent into an analog-to-digital converter having a 12 bit resolution without loss of information.

A resolution of 12 bits means that digital data can vary between zero/one and 4095/4096 (4096 equals 2 to the power of 12; hence 12 bit resolution). With respect to the original chromatographic signal, 12 bit resolution means that analog signals between 10 microvolts and 40.96 millivolts (level one) or 2 millivolts and 8.2 volts (level two) could be digitized without any dubious compromise at an affordable level of investment in hardware tools. The price of A/D converters increases exponentially with the resolution.

Most of our chromatographic analysis is done within one, and rarely within more than three orders of magnitudes with respect to time and concentration (20 seconds is a short elution time, but 20000 seconds is a working day!).

Ppb quantitation is accomplished by enrichment, not by any six order of magnitude linearity of an electron capture detector! Nevertheless, some GC analyses may be performed, in cases of multilevel calibration, in a range encompassing six orders of magnitude. And for this we use a 12 bit resolution A/D converter and a software-switchable signal amplifier (or signal conditioner).

12 bit data can easily be handled as INTEGER values. Integer data can be stored at very high density on floppy disks or into memory (RAM) or on hard disks. Two bytes of storage capacity are used to store one 12 bit value. One floppy (5.25 inch) can hold 65000 raw data single sided; one hard disk of 21 megabyte can hold 120 disks full of 12 bit raw data. Details of raw data storage are discussed below in Section 3.9.

3.7 Software Necessary

P.K. Warme [2] has written an interface board programming language which makes it easy to program timing events, analog-to-digital conversions, digital-to-analog conversions, data storage, data retrieval by using a 12 bit AD/DA converter board of the "IMI"-type [2].

We will present a detailed discussion of interface programming in chromatography in a later volume.

For the time being we shall be content to consider only the key algorithms and commands in order to demonstrate the ease and elegance of programming raw data measurement and storage in BASIC.

The following little program does the trick:

Procedures to initialize the raw data measurement:
> load the machine language
> load the machine program which starts clock, timers, AD conversion

> prepare the control of data flow, for instance
    data number
    time
    signal value
onto the computer screen as digital output
    or as graphics display
ASK for "routine" or "special run"
    IF routine, THEN install routine run data and GOTO START,
ELSE GOTO DIALOG for accepting
    number of data to measure
    conversion rate or
    elapsing time for measurement

DIALOG (to define measurement conditions):
    AD = amount of data to take
    DT = time between measurements
> START the measurements (conversion, storage) by a computer/analyst dialog or just a wait signal ("until key press")
and after the start happened:

```
100 FOR I = 1 TO AD
110 T = T - DT
120 & TI1 : IF D% > T GOTO 120
```

REMARK: in the following we will have REMARKS, this is for the reader and REM's which stand in the program lines. Here a REMARK: D% is the time called from the on-board clock. "& TI1" sends the time value. DT is the time interval from conversion to conversion. Line 120 makes sure that only at each time slice of DT seconds the conversion happens in line 130

```
130 & AI1,I
```

REMARK: "& AI1" "does it". It calls the machine language part commanding: "now take the analog value and convert it into digital form. Give it the data number 'I', store the digital value as D%(I) in the computer RAM".

140 GOSUB 1000 : REM (PRESENTATION)

REMARK: "presentation" is a subprogram which transforms the digital measurement into a point plotted graphically whose height corresponds to the signal intensity and whose x-axis position corresponds to the time elapsed since start.

150 PRINT DT*I, D%(I)*S : NEXT

REMARK: DT*I represents the time elapsed in seconds; D%(I) is the digital value of the analog signal; S is the factor which converts the digital number back into the original analog signal.

In a practical example, the 12 bit conversion range might be switched to 500 mV. Then 500 (analog) mV corresponds to 4096 (digital) and in this case S = 500/4096. In other words, 0 bit = 0 mV and 12 bit full range (= 4096) corresponds to 500 mV, and any intermediate value is transformed back into the analog signal value on multiplication by the factor S.

160 GOSUB 2000 :

REMARK: RAW DATA STORAGE or in case of non-automatic measurement: DIALOG for storage of the digital raw data onto DISK or somewhere else, asking for the raw data set NAME *etc.*

```
170 REM DIALOG
175 INPUT "any next analysis (YES:NO)" ; Q$
180 IF yes GOTO START
190 END
1000 REM subprogram screen data in graphics
1100 RETURN
2000 REM subprogram data storage
2010 INPUT "enter instr.number";NA$
```

REMARK: the instr. number can be taken automatically from the interface, the date is given in numbers of years and days, and the numbers of hours and minutes could have been stored by a clock on-board. Let us assume that these data are named DA and stored in the memory, then the NAME of the to be stored data set is automatically available and it follows:

```
2030 NA$ = "G"; DA + NA$ + "/";S + "/";DT
2040 PRINT CHR$(4);"OPEN";NA$
2050 PRINT CHR$(4);"WRITE";NA$
2060 FOR I = 1 TO AD
2070 PRINT D%(I)
2080 NEXT
2090 PRINT CHR$(4);"CLOSE"
2100 RETURN
```

The key part of the software program "raw data digitization and storage" is in lines 100 — 150 (measurements) and 160 (storage) together with the subprograms which handle details.

3.8 Raw Data Naming Convention

Let us assume that we want to store 2000 raw data on a floppy disk. We must give the data set a name. This name might be

NA$ = G(date-instrument/S/MT)

By G we define the procedure.
By "date-instrument" we provide a full documentation of WHAT, WHERE, HOW, WHO, which includes ALL details of the sample because at a given moment of the day ONE instrument can take only ONE sample. Hence sample and "date-instrument" are strictly correlated.

Of course, in order to translate the "date-instrument" code into SAMPLE (who took it how and when from where into what container, and who took a secondary sample how into what, treating it according to which procedure..., and who took the final analytical tertiary sample to inject it under what conditions, how, into the GC equipment number), this information must be decoded from a data base program. We will discuss "naming conventions and automatic lab data organization" in a subsequent volume.

"Date" can be coded as DAY of the year and MINUTE of the day. Varian Ass. use this principle in their chromatography data stations. A number such as 1....365/0......1440 is easy to understand and simple to handle automatically by lab organization programs.

Now let us assume you are working in a big company with a maximum of 9999 gas chromatography instruments, analyzing a maximum of 9999 x 1440 = 14 million differing sample types a day—at least this is the "capacity" of the proposed coding of data set names.

However, there is even more to the name NA$, as given above:

NA$ = G224.1009-9885/.12/.15 contains absolutely all the details one might want to store about an analysis. It also contains all details of sample taking, sample preparation, and sample treatment, as well as all details about the chromatographic equipment, the detector, the column, the temperatures, sensitivities, and even details of any defects or analytical anomalies. This is because the specific overall conditions of an instrument used and the procedure adopted can be coded in full JUST by time. The only prerequisite is a data base program which is TIME related. We will discuss this in a subsequent volume. The sample was product "G224.1009-9885", because into instrument no. 9885G (a gas chromatograph as G is used) one could inject on the 224th day of the year (July 12 in 1983) only ONE sample at time 16:49. At this minute it was the unique sample, coded by time and instrument number: the sample no. "G224.1009-9885". It would be redundant to use any more code in the name of the raw data set.

By means of the additional information ".../.12/.12" we are able to reconstruct the analog chromatogram from the stored chromatogram raw data in all details:
The "/.12/" says that the AD converter used a 500 mV full scale range. A digital value 409 therefore would have represented a 50 mV signal height. If this was the 985th value, then the 50 mV signal corresponds to the elution time 985 * .12, which was 118.2 seconds after the start of analysis. This is because .12 was the conversion rato, expressed in seconds per single measurement.

The result is plain to see: total documentation without any loss of information. Computer chromatography will dramatically change lab data management and lab organization towards a very high level of protection against disorder.

The raw data are now stored magnetically on a floppy disk. The name of the raw data set is also stored on the same disk. Data and name are correlated by the Disk Organization System DOS which is available as software in the computer after it has been started.

3.9 Storage Capacity of Raw Data on Disk

Using 2 byte INTEGER raw data storage principles, one can store, at the time of writing of the present volume, up to 240 raw data sets on both sides of a floppy disk.

This holds true as long as the data sets are no larger than 1024 raw data (theoretical maximum: 240 chromatograms of below 1024 raw data points each per floppy). About 100 large data sets of about 2000 raw data each can be stored on a double sided floppy disk, and only four extra large data sets of 32000 raw data each can be stored per floppy disk using storage techniques available today. New developments will improve this situation remarkably in the near future.

NOTE: only the digitalized chromatographic signal is stored on disk. The corresponding TIME is coded in the serial NUMBER of the piece of raw data .

We will discuss storage and retrieval details later. At the present juncture, it will be interesting to consider an example of simple data "playback" onto the computer display screen as follows.

3.10 Raw Data Retrieval

With the following BASIC program, raw data can be read and translated into chromatogram values, giving times in seconds and signal heights in mV:

```
100 REM raw data retrieval
110 INPUT " ENTER NAME OF DATA SET ";NA$
120 IF NA$ = "?" THEN PRINT
CHR$(4);"CATALOG";:PRINT
    :GOTO 110
130 IF NA$ = "/" THEN END
140 S = VAL (RIGHT$ (NA$,7))
150 DT = VAL (RIGHT$ (NA$,3))
160 PRINT CHR$(4);"OPEN";NA$
170 PRINT CHR$(4);"READ";NA$
180 INPUT AD
190 FOR I = 1 TO AD − 1
200 INPUT D%(I)
210 W%(I) = D%(I) * S
```

REMARK: W%(I) are now the original analog values

```
230 NEXT : PRINT CHR$(4);"CLOSE"
```

REMARK: the display of the recalculated values and the time of each measure raw data point are printed onto the screen by the statements in line 250:

```
250 FOR I = 1 TO AD − 1
260 PRINT I, I * DT, INT(10 * W%(I) + .5)/10
270 NEXT
```

REMARK: the "INT(10 * W%(I) +.5)/10" algorithm makes

sure that the analog data are displayed in .1 mV units without rounding error. I $*$ DT recalculates the time corresponding to the signal. The "," between the PRINT statements makes sure that the display is given in three rows on the screen, numbered by I.

An experimental raw data set—condensed in this case to just 6 rows—is given in Table (1). It illustrates the enormous amount of data comprising just a short chromatogram. Lines of asterisks ("*****") locate peaks. The reader will experience difficulty in detecting the peaks without indicators. (Try to find some trace peaks!). The signals are weak: the maximum peak height in Table (1) was detected at measurement no. 344 with a signal height of 517 (digital) representing 63 mV at maximum sensitivity of the detector/amplifier (GC, FID, fused silica, split ratio 1:250, 0.1% test concentration in solvent *n*-undecane).

It is pointless for the analyst to examine raw data; the computer does this very much faster and much more efficiently in every respect. Nevertheless: You might want to see a full data set just to gain an idea of the quantity or to appreciate various details . Here it is:

Table (1)
A full raw data set: 2000 values, 6 larger peaks, maxima at **. This material is for rechecking by interested readers. It represents a "bad" capillary test gas chromatogram. See baseline shift in Figs. (7,8).**

NR	SIG	SIG	SIG	SIG	SIG	SIG	SIG	SIG
1 - 8	8	8	7	7	7	7	7	7
9 - 16	7	7	7	8	7	8	0	8
17 - 24	8	8	8	8	8	7	7	8
25 - 32	7	7	7	7	8	8	8	8
33 - 40	8	8	8	8	8	8	8	8
41 - 48	8	8	8	8	8	8	8	8
49 - 56	9	8	9	9	9	9	9	8
57 - 64	9	8	8	8	8	8	9	8
65 - 72	9	9	9	9	9	9	9	9
73 - 80	9	9	9	9	9	9	9	9
81 - 88	9	9	9	10	10	10	10	10
89 - 96	10	10	10	10	10	10	10	10
97 - 104	11	11	11	11	11	11	11	11
105 - 112	11	12	12	11	12	11	12	12
113 - 120	12	12	13	12	12	12	13	13
121 - 128	13	13	12	12	13	13	13	13
129 - 136	13	13	13	13	13	14	13	14
137 - 144	13	13	14	13	13	14	14	14
145 - 152	14	14	14	14	14	14	14	14
153 - 160	14	14	14	13	14	14	14	14

Table 1 GC raw data set

161 - 168	14	14	14	14	14	14	14	14	
169 - 176	14	14	14	13	13	13	13	13	
177 - 184	14	14	14	14	14	14	14	14	
185 - 192	14	14	14	14	14	14	14	14	
193 - 200	14	14	14	14	14	14	14	14	
201 - 208	14	14	14	14	14	14	14	14	
209 - 216	14	14	14	14	14	14	14	14	
217 - 224	14	15	15	14	14	14	15	15	
225 - 232	15	14	15	15	15	15	14	15	
233 - 240	15	15	14	15	15	15	15	15	
241 - 248	15	16	16	15	15	15	14	14	
249 - 256	15	14	14	14	15	15	16	15	
257 - 264	15	15	15	15	15	15	15	15	
265 - 272	15	15	15	15	15	14	15	15	
273 - 280	15	15	15	15	15	15	15	15	
281 - 288	15	15	15	15	16	16	15	15	
289 - 296	15	15	15	15	15	15	15	15	
297 - 304	15	15	15	15	15	15	15	15	
305 - 312	15	15	15	15	15	15	15	15	
313 - 320	15	15	15	15	15	15	16	16	
321 - 328	17	17	16	16	16	16	16	16	
329 - 336	15	15	15	15	15	15	15	15	
337 - 344	16	15	15	15	19	44	150	366	*****
345 - 352	517	454	245	110	43	24	19	17	
353 - 360	17	16	16	16	15	16	15	15	
361 - 368	15	15	15	15	15	15	15	15	
369 - 376	15	15	15	15	15	16	15	15	
377 - 384	16	15	15	15	15	15	15	15	
385 - 392	15	15	15	15	15	15	15	15	
393 - 400	15	15	15	15	15	15	15	15	
401 - 408	15	15	15	15	16	16	16	16	
409 - 416	16	16	16	15	16	15	16	16	
417 - 424	15	15	15	15	15	15	16	16	
425 - 432	15	15	16	16	16	15	16	16	
433 - 440	16	15	15	15	15	15	15	15	
441 - 448	15	15	15	16	15	16	16	16	
449 - 456	16	16	16	17	19	32	68	157	
457 - 464	286	379	395	306	195	99	49	29	*****
465 - 472	23	19	18	17	16	16	16	15	
473 - 480	15	15	15	15	15	15	15	15	
481 - 488	15	15	15	15	16	16	16	16	
489 - 496	16	16	16	16	16	16	16	16	
497 - 504	16	16	16	16	15	15	15	15	
505 - 512	15	15	15	16	16	16	16	16	
513 - 520	16	16	16	16	16	15	15	15	
521 - 528	15	15	15	15	15	15	15	15	
529 - 536	16	15	16	16	16	16	16	16	
537 - 544	16	16	16	16	16	15	16	15	
545 - 552	16	15	15	15	15	15	15	15	
553 - 560	16	16	16	16	16	16	16	16	

Table 1 GC raw data set

561 - 568	16	16	16	16	15	15	15	15	
569 - 576	15	16	16	16	16	17	16	17	
577 - 584	16	16	16	16	16	16	16	16	
585 - 592	16	16	16	16	15	15	15	16	
593 - 600	16	16	16	16	17	18	18	19	
601 - 608	19	18	17	16	16	16	15	15	
609 - 616	15	15	15	15	15	16	15	15	
617 - 624	16	16	16	16	16	16	16	16	
625 - 632	16	15	16	15	15	15	15	15	
633 - 640	15	16	15	15	15	15	15	15	
641 - 648	15	15	16	16	16	16	16	16	
649 - 656	16	16	16	15	16	15	15	15	
657 - 664	15	15	15	16	15	15	16	16	
665 - 672	16	17	19	25	39	67	109	177	
673 - 680	254	325	365	364	313	244	165	104	*****
681 - 688	68	43	30	23	20	19	18	17	
689 - 696	17	16	16	15	16	15	16	15	
697 - 704	15	15	15	15	15	15	15	16	
705 - 712	16	16	16	16	16	16	15	15	
713 - 720	15	15	15	15	15	14	14	14	
721 - 728	14	14	14	14	14	13	14	14	
729 - 736	13	13	13	12	12	12	12	11	
737 - 744	11	11	10	10	10	10	10	10	
745 - 752	9	9	9	9	9	9	9	8	
753 - 760	8	8	8	8	8	8	7	7	
761 - 768	7	8	8	8	8	8	8	8	
769 - 776	8	8	8	8	8	8	7	8	
777 - 784	8	8	8	7	8	8	8	8	
785 - 792	8	8	8	8	8	8	8	8	
793 - 800	8	8	7	8	8	8	8	8	
801 - 808	8	8	8	8	8	8	8	8	
809 - 816	8	8	0	0	8	8	8	7	
817 - 824	8	8	8	8	8	8	8	8	
825 - 832	7	8	8	7	8	8	8	8	
833 - 840	8	8	8	8	8	8	8	8	
841 - 848	8	8	8	8	8	8	8	8	
849 - 856	8	8	8	8	8	8	8	8	
857 - 864	8	8	8	8	7	7	7	7	
865 - 872	8	8	8	8	8	8	8	8	
873 - 880	8	8	8	7	8	8	8	8	
881 - 888	8	8	8	8	8	8	8	8	
889 - 896	8	8	8	8	8	8	8	7	
897 - 904	8	8	8	8	8	8	8	8	
905 - 912	8	8	8	8	8	8	8	8	
913 - 920	8	8	8	8	8	7	7	7	
921 - 928	8	7	8	8	8	8	8	8	
929 - 936	8	8	8	8	8	7	8	8	
937 - 944	7	7	8	8	8	8	8	8	
945 - 952	8	8	8	8	8	8	8	8	
953 - 960	8	8	8	7	7	8	8	8	

Table 1 GC raw data set

961 - 968	8	8	8	8	7	8	8	8	
969 - 976	8	8	8	8	8	8	8	8	
977 - 984	8	8	8	8	8	8	8	8	
985 - 992	8	8	8	8	8	8	8	8	
993 - 1000	8	8	8	8	7	8	8	8	
1001- 1008	8	8	8	8	8	8	8	8	
1009- 1016	8	8	8	8	8	8	8	8	
1017- 1024	8	8	7	7	8	8	8	8	
1025- 1032	8	8	8	8	8	8	8	8	
1033- 1040	8	8	8	8	8	7	8	8	
1041- 1048	8	8	8	8	8	8	8	8	
1049- 1056	8	8	8	8	8	8	8	8	
1057- 1064	8	7	7	7	8	8	7	8	
1065- 1072	8	8	8	8	8	8	8	8	
1073- 1080	8	9	10	12	16	22	32	46	
1081- 1088	63	89	122	161	197	236	271	293	
1089- 1096	303	296	276	242	201	158	122	88	*****
1097- 1104	62	45	32	23	18	14	12	10	
1105- 1112	10	9	9	8	8	8	8	7	
1113- 1120	7	7	7	7	8	8	8	8	
1121- 1128	8	8	8	7	8	8	8	8	
1129- 1136	7	8	8	8	8	8	8	8	
1137- 1144	8	8	8	8	8	8	8	7	
1145- 1152	8	8	8	8	8	8	8	8	
1153- 1160	8	8	8	8	8	8	8	8	
1161- 1168	8	8	8	8	8	8	8	8	
1169- 1176	8	7	8	8	8	8	8	8	
1177- 1184	8	8	8	8	8	8	8	8	
1185- 1192	8	7	8	8	8	8	8	8	
1193- 1200	8	7	8	8	8	7	7	8	
1201- 1208	8	8	8	8	8	8	8	8	
1209- 1216	7	7	7	8	8	8	8	8	
1217- 1224	8	8	8	8	8	8	8	8	
1225- 1232	7	7	7	8	8	8	8	8	
1233- 1240	8	8	8	8	8	8	8	8	
1241- 1248	8	7	8	8	8	8	8	8	
1249- 1256	8	8	8	8	8	8	7	7	
1257- 1264	7	8	8	8	8	8	8	8	
1265- 1272	8	8	8	7	7	7	8	8	
1273- 1280	8	8	8	8	8	7	8	8	
1281- 1288	8	7	7	7	7	8	8	8	
1289- 1296	8	8	8	8	8	7	8	8	
1297- 1304	8	7	7	7	8	8	8	8	
1305- 1312	8	8	8	8	8	8	8	7	
1313- 1320	8	8	8	8	8	8	8	8	
1321- 1328	8	8	7	7	8	8	8	8	
1329- 1336	8	8	8	8	8	8	8	8	
1337- 1344	7	8	8	8	8	8	8	8	
1345- 1352	8	8	8	8	7	8	8	7	
1353- 1360	8	8	8	8	8	8	8	8	

Table 1 GC raw data set

1361- 1368	8	8	8	8	8	8	8	8	
1369- 1376	8	8	8	8	8	8	8	8	
1377- 1384	8	8	8	8	8	8	8	8	
1385- 1392	8	8	8	8	8	8	8	8	
1393- 1400	8	8	8	8	8	8	8	8	
1401- 1408	8	8	8	8	8	8	8	8	
1409- 1416	8	8	8	8	8	8	8	8	
1417- 1424	8	8	8	7	8	8	8	8	
1425- 1432	8	8	8	8	8	8	8	8	
1433- 1440	8	8	8	8	8	8	8	8	
1441- 1448	8	8	8	8	8	8	8	8	
1449- 1456	8	8	8	8	8	8	8	8	
1457- 1464	8	8	8	8	8	8	8	8	
1465- 1472	8	9	9	9	9	9	8	8	
1473- 1480	8	8	8	8	8	8	8	8	
1481- 1488	8	8	8	8	8	8	8	8	
1489- 1496	8	8	8	8	8	8	8	8	
1497- 1504	8	8	8	8	8	8	8	8	
1505- 1512	8	8	8	8	8	8	8	8	
1513- 1520	8	8	8	8	8	8	8	8	
1521- 1528	8	8	8	8	9	8	8	8	
1529- 1536	8	8	8	8	8	8	8	8	
1537- 1544	8	8	8	8	8	8	8	8	
1545- 1552	8	8	8	8	8	9	8	8	
1553- 1560	8	8	8	8	8	8	8	8	
1561- 1568	8	8	8	8	8	8	8	8	
1569- 1576	8	8	8	8	8	8	8	8	
1577- 1584	8	9	8	9	9	9	9	9	
1585- 1592	9	9	9	10	10	10	10	10	*****
1593- 1600	10	10	10	9	9	9	9	8	
1601- 1608	9	9	8	9	9	8	8	8	
1609- 1616	8	8	8	8	8	8	8	8	
1617- 1624	8	8	8	8	8	8	7	7	
1625- 1632	7	8	8	8	7	8	8	8	
1633- 1640	8	8	8	8	8	7	7	7	
1641- 1648	7	8	8	8	8	8	8	8	
1649- 1656	8	8	7	7	7	7	8	8	
1657- 1664	8	8	8	8	8	8	8	8	
1665- 1672	8	7	8	7	7	8	8	8	
1673- 1680	8	8	8	8	8	8	8	8	
1681- 1688	7	8	8	8	8	8	8	8	
1689- 1696	8	8	7	7	7	8	7	7	
1697- 1704	8	8	8	8	8	8	8	8	
1705- 1712	8	7	7	7	8	8	7	8	
1713- 1720	8	8	8	8	8	8	8	8	
1721- 1728	8	8	7	8	8	8	8	8	
1729- 1736	8	8	8	8	8	8	8	8	
1737- 1744	8	8	8	8	8	8	8	8	
1745- 1752	8	8	8	8	8	8	8	8	
1753- 1760	8	8	8	8	9	9	9	9	

Table 1 GC raw data set

1761- 1768	9	8	8	8	8	8	8	8	
1769- 1776	8	8	8	8	8	8	8	8	
1777- 1784	8	8	8	8	8	8	8	8	
1785- 1792	8	8	8	8	8	8	8	8	
1793- 1800	8	8	8	8	8	8	8	8	
1801- 1808	8	8	8	8	8	8	8	8	
1809- 1816	8	8	8	8	8	8	8	8	
1817- 1824	8	8	8	8	8	8	8	8	
1825- 1832	8	8	8	8	8	8	8	8	
1833- 1840	8	8	8	8	8	8	8	8	
1841- 1848	8	8	8	8	8	8	8	9	
1849- 1856	9	9	10	11	12	14	17	20	
1857- 1864	23	27	33	41	49	60	72	87	
1865- 1872	101	119	138	153	172	191	208	228	
1873- 1880	248	268	284	300	314	323	329	328	*****
1881- 1888	323	313	297	275	251	222	192	166	
1889- 1896	138	112	93	73	57	45	37	29	
1897- 1904	23	19	16	14	12	11	10	9	
1905- 1912	9	9	9	9	9	9	8	8	
1913- 1920	8	8	8	8	8	8	8	8	
1921- 1928	8	8	8	8	8	8	8	8	
1929- 1936	7	8	8	8	8	8	8	8	
1937- 1944	8	8	8	8	8	8	8	8	
1945- 1952	8	8	9	9	9	9	9	9	
1953- 1960	9	9	9	9	10	10	10	10	
1961- 1968	10	10	11	11	11	10	11	10	
1969- 1976	10	11	10	10	11	10	10	10	*****
1977- 1984	9	9	9	9	9	9	9	9	
1985- 1992	9	9	9	9	9	9	8	8	
1993- 2000	8	8	8	8	8	8	0	0	

DATAFILE NAME : DA1

MAXIMUM = 517 AT T = NR 345 MINIMUM = 7 AT T = NR 3

4 Experimental

Digitization, raw data storage, raw data analysis, and concentration down to chromatography-relevant values are discussed in depth for two examples.

The raw data used are given with all details and in full length for readers interested in numbers. Peak maxima are found by higher order polynomial regression or the cubic spline technique. It is shown that
a) time data cannot be precise enough
b) the question of which time value is the 'correct' one can be answered: by cross correlation.

Practical and theoretical peak maxima are compared and the precision and accuracy of results gained by differing mathematical models are shown in data tables. As there is no difference between gas and high pressure liquid chromatography, both techniques are treated simultaneously in parts of this chapter.

When it comes to DEAD TIME calculation, however, GC differs from HPLC and here the treatment diverges. Which dead time calculation procedure is "the better" one or which value is falsified or correct can be checked again by data cross correlation.

The use of a lab computer in data concentration offers most valuable additional information: about systematical errors. It is discussed in detail how this might help the chromatographer in practice.

Concentrate

From raw data measurement, through storage, concentration, and reduction, two sets of data are used: 2000 measurements for GC, 1000 for HPLC. Procedures to find the best values for retention time and dead time in GC and HPLC are discussed, together with the principles of testing for systematically wrong or the best acceptable results.

4.1 General

The hardware (chromatographic equipment, computer, interfaces, *etc.*) are defined in the **Appendix** under Hardware & Software/On-Line Chromatograph.

- We injected mixtures of homologous series into the GC or HPLC instruments.
- We ran the chromatogram.
- We stored the raw data on disk (see the Section on Raw Data Storage).
- We analyzed the raw data in steps as follows:

a) Replot raw data onto screen.

b) Cut with the aid of cursor positioning those parts of the data which belong to one single peak; or do the same by using an automatic program—see under "Autovideogration".

c) Find the largest peak height value within each peak data group; call this value the raw peak maximum; take 3 points to the left, and three points to the right of this raw maximum value. Table (3) (p. 57) shows a result.

d) Calculate by regression the "theoretical" maximum using 2nd, 3rd, 4th, and 5th order polynomial least-squares interpolation, or, if the data do not allow least-squares techniques, then higher order INTERPOLATION is used to optimize the peak maximum value. The difference between the former and the latter is seen in Figure (5), a 2nd order polynomial fit through the three top peak raw data is seen in Figure (4) and was used almost always in the past.

e) Find the baseline. The perpendicular from the "theoretical maximum" to the baseline is the theoretical peak height. This distance is often greater than the maximum peak height signal found from raw data.

f) Calculate half of the peak maximum height.

g) Find two raw data points above peak half height and below peak half height to the LEFT of the peak maximum.

h) Repeat procedure g) to the RIGHT of the peak maximum.

i) Use linear regression through the 4 raw data points to the left and through the 4 raw data points to the right of the peak maximum.

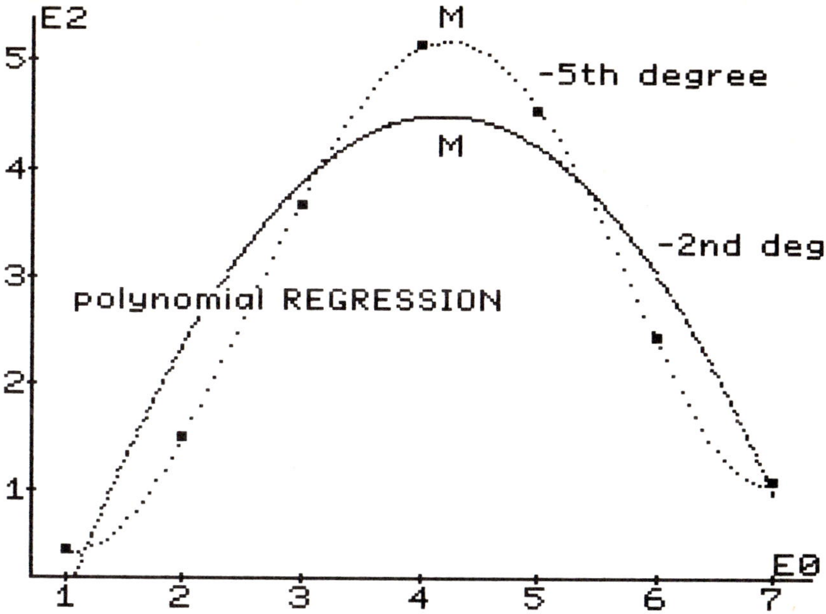

Figure (4)
Raw data, peak 1 of the set in Table (1). 2nd and 5th degree polynomial regression is used to find the retention time. 2nd order regression would NOT find the peak height. Noisy points: see Fig. (5).

j) Calculate the linear regression factors "left" and "right" as follows:

$$YL = AL * t + BL \qquad YR = AR * t + BR$$

(t = time axis data)

k) Now cut these two lines with the parallel to the time axis at half peak height which has the formula

$$HH = 0 * t + BH$$

HH is half the peak height.
Three straight lines meet at two points with the coordinates height = HH ; time = t(L); and height = HH; time = t(R).

The peak width at half height finally is

$$t(R) - t(L) = b\,(.5)$$

This value can now be calculated according to
YL = YR = HH (height position of the intercept of the three linear functions).

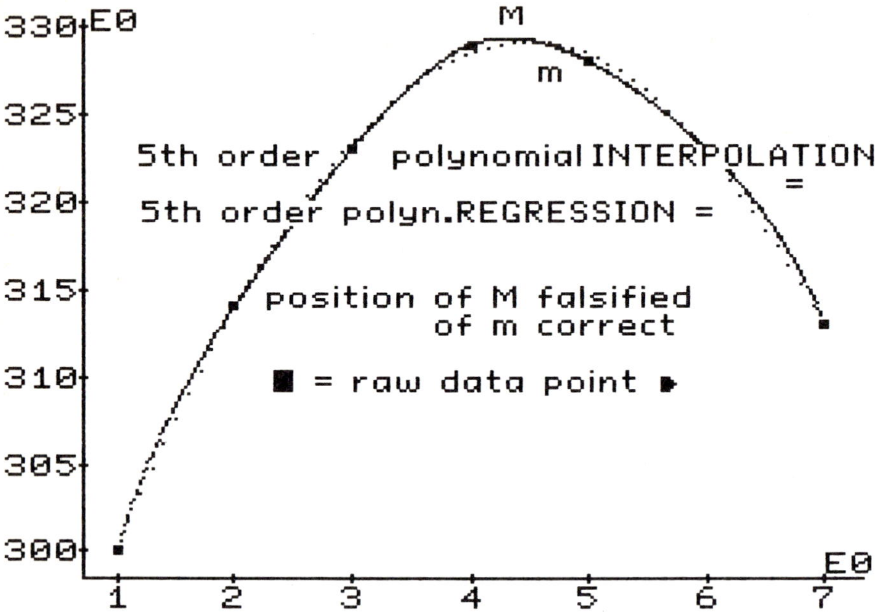

Figure (5)
Raw data, peak 5 of the set in Table (1). 5th order polynomial INTERPOLATION versus REGRESSION. In case of very noisy raw data "regression" must be used. See influence on position of m versus M, which determines "tms".

$$AL * t(L) + BL = AR * t(R) + BR$$

$$t(R) - t(L) = (HH - BR)/AR - (HH - BL)/AL$$

It is evident that an inaccurate peak maximum value affects all other time axis dependent data.

The existence of NOISE has to be taken into account. Therefore the utmost care must be exercised in making each digital raw data point as safe as possible. This approach is superior to digital smoothing techniques executed with the raw data set.

The use of the most intelligent methods to achieve accurate raw data measurements is a must. Success in this early task of computer chromatography allows application of clever data handling procedures instead of just noise suppression algorithms which change the true peak shape and the true time data.

This all has to be taken into account prior to selection of a digitization technique.

Finally, we have to digitize because our computers need digital data. Analog computing procedures are not very popular, and digitization is therefore performed in almost all data handling systems.

Integrators digitize the analog signal. They use up to 40 smoothed data per second and put single data points together after digital or analog or a combined procedure of further smoothing. They concentrate data and "kill" the original data. If, during the concentration process, the signal structure has been altered, then the truth has "vanished".

Finally, almost all existing integrators store and print immediately, or print after completion of analysis, the peak raw retention time value to within a precision of one full or one tenth of a second. Many newer integrators including some computer integrators offer a retention time data precision of no better than 0.01 minute = 600 milliseconds. Very few of the most recent computerized data terminals are able to print out peak retention time data to 50 or 20 milliseconds precision. However, for successful cross correlation of data we need 10 millisecond precision together with 10 millisecond absolute accuracy.

On use of a recorder to draw a chromatogram of, say, 10 minutes' running time on a sheet of paper 1000 mm long, one could interpolate retention time data at a sharp peak maximum to within about 0.5 mm local precision. This represents a 300 millisecond precision and costs a lot of working time. Mechanical care as well as an unshakable belief in the constancy of paper feed rate is needed. (Do not believe in that: the paper transporting systems are often dependent upon the line frequency, and the line frequency may vary considerably within short periods! It only looks constant, as the line frequency is regulated in the power stations to meet the requirements for a constant 24 hour cycle.)

This manual procedure might appear to be an analog one, but is actually digital due to the limits of length measurement precision. Nevertheless, the time consuming paper handling technique described here can afford quite satisfactory results if used in conjunction with a very precise clock. Precise high speed recorder data run at constant frequency and checked by a digital clock allowed us to pinpoint a lot of computer integrator ailments with respect to time scaling. When the results were reported, the author was told that he was unable to make correct measurements—but that's life. Analytical data, if correct, are not always the most dearly beloved of data, especially when they contradict what one believes or (worse) trusts. There is a fair amount of inaccuracy on the market, at high prices and bearing highly respected names! There are incredible systematic errors hardly anybody would believe, as we all are in danger of becoming blind believers in digital data!

Figure (6) shows the schematic of the instrument, Table (1) the total of raw data of a test chromatogram in GC, Figure (7) the chromatogram as visible on screen (with very poor resolution!!), and Figure (8) as reproduced from raw data by a laboratory recorder on-line D/A converter (back conversion).

Table (2) shows HPLC raw data of 5 homologous test peaks, and Table (3) the concentrate of a time data concentrate prepared for a cross correlation analysis of retention time values.

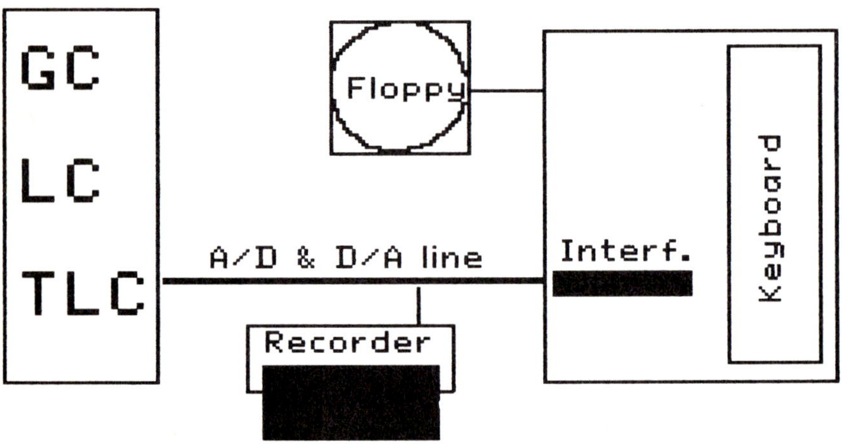

Figure (6)
The analog signal from a GC or LC instrument or a scanner is fed into the A/D converter. The raw data are stored on a floppy. Through the D/A converter the reconstructed analog signal is fed into a recorder, or integrator, or used as "IF THEN" control value fed back into parts of the chromatographic instruments to adjust flow, temperature, column selection, detector data, sensitivity, sample no., etc.

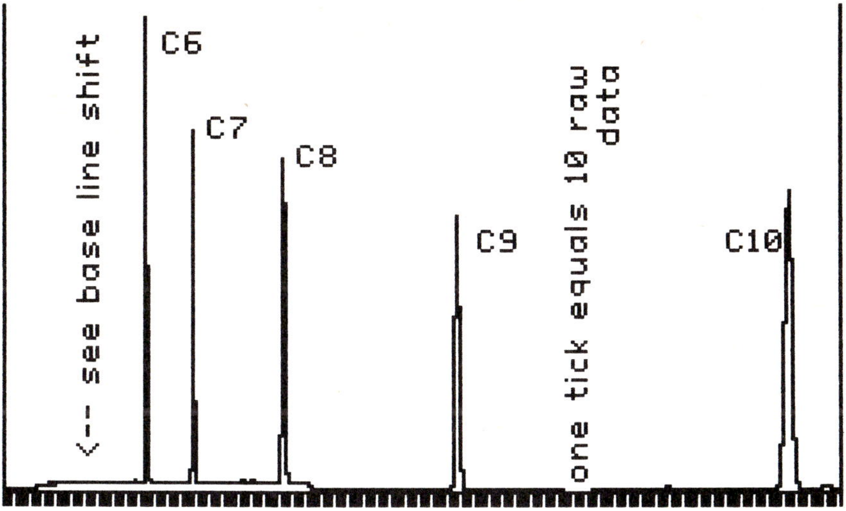

Figure (7)
Graphics display of a raw data set (Table 1 data) on a screen looks very bad. The impression is that 12 bit raw data can never reconstruct the well known good chromatogram precision of 0.03% at fsd. The screen resolution of simple micros is only 1 in 270.

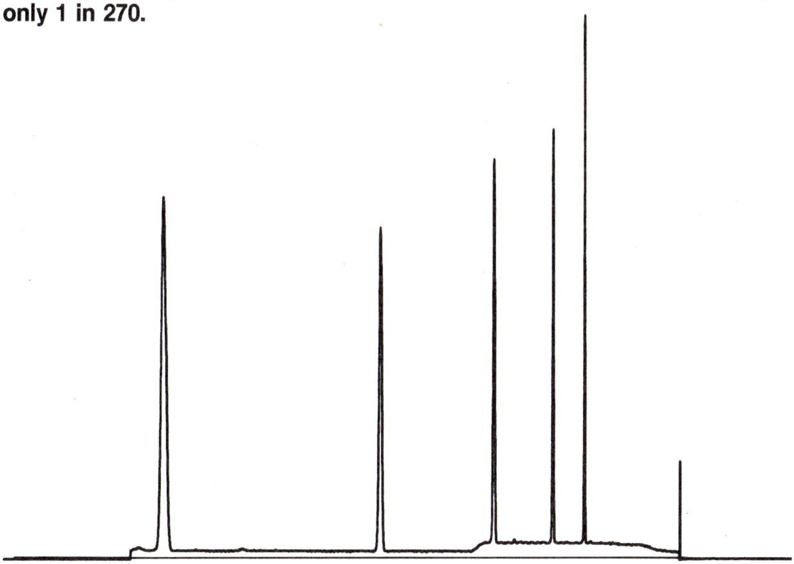

Figure (8)
"Playback" of raw data, reconverted into the analog signal using the D/A feature of an AD/DA interface: the digital signal has been recalculated so that the largest peak analog signal has the fsd value. 12 bit resolution; 500 mV fsd; playback rate: 0.1 s per raw data point. Recorders time constant: 100 ms for fsd.

Table (2): HPLC raw data set of 5 test peaks.

NR	SIG	SIG	SIG	SIG	SIG	SIG		
1 - 8	3	4	4	4	5	5	5	5
9 - 16	5	5	5	5	5	4	4	4
17 - 24	4	4	4	4	4	4	4	4
25 - 32	4	4	4	5	4	4	4	4
33 - 40	5	5	5	5	5	5	5	5
41 - 48	5	5	5	5	5	5	4	5
49 - 56	5	5	5	5	5	5	5	4
57 - 64	5	4	4	4	4	3	3	3
65 - 72	3	3	4	4	4	4	4	4
73 - 80	5	5	5	6	6	8	10	9
81 - 88	8	7	7	6	6	5	5	5
89 - 96	5	5	5	4	5	4	4	4
97 - 104	1	5	9	19	30	36	36	21
105 - 112	4	-6	-9	-9	-6	-3	1	4
113 - 120	6	5	4	5	6	9	13	17
121 - 128	19	22	24	27	32	38	45	49
129 - 136	53	54	55	53	52	49	46	44
137 - 144	41	37	33	30	26	23	19	16
145 - 152	14	13	12	12	14	15	16	17
153 - 160	16	14	12	10	9	9	10	12
161 - 168	16	20	24	25	26	28	35	48
169 - 176	59	61	54	42	30	22	16	13
177 - 184	12	11	10	9	9	8	9	9
185 - 192	11	13	16	20	27	37	46	49
193 - 200	45	36	27	20	15	12	10	9
201 - 208	8	7	7	7	7	8	7	7
209 - 216	7	8	8	10	12	14	16	16
217 - 224	15	14	12	10	10	9	9	8
225 - 232	8	7	7	6	6	8	12	27
233 - 240	58	112	186	258	303	305	269	215
241 - 248	163	121	92	72	59	50	44	38
249 - 256	33	29	25	21	18	15	13	11
257 - 264	10	9	8	8	8	8	8	10
265 - 272	10	12	11	11	10	9	8	8
273 - 280	7	6	6	6	7	8	13	24
281 - 288	45	83	137	203	269	316	332	313
289 - 296	269	215	163	121	90	69	53	43
297 - 304	35	29	25	21	19	16	14	13
305 - 312	12	10	9	9	8	8	7	7
313 - 320	6	6	6	6	6	6	6	6
321 - 328	5	6	5	6	6	6	6	6
329 - 336	6	6	7	7	7	7	7	7
337 - 344	6	6	6	6	6	6	7	9
345 - 352	11	16	24	37	58	87	124	166
353 - 360	207	240	257	257	239	209	175	140

Table (2): HPLC raw data set

361 - 368	110	86	69	57	48	42	38	34
369 - 376	31	28	25	22	19	17	15	13
377 - 384	12	10	9	9	8	8	7	7
385 - 392	7	6	6	6	6	6	5	6
393 - 400	6	6	6	6	6	6	5	5
401 - 408	5	5	5	5	5	5	5	5
409 - 416	5	5	5	5	5	5	5	5
417 - 424	5	5	5	5	5	5	5	5
425 - 432	5	5	5	5	5	5	5	5
433 - 440	5	5	5	5	5	5	6	7
441 - 448	8	11	16	24	36	54	77	107
449 - 456	139	174	209	239	262	274	274	263
457 - 464	241	213	183	152	125	101	82	67
465 - 472	56	47	41	36	32	29	26	23
473 - 480	22	19	18	16	14	13	12	11
481 - 488	10	9	8	8	7	7	7	7
489 - 496	7	6	6	6	6	6	5	5
497 - 504	5	5	5	4	5	4	5	5
505 - 512	5	5	5	5	5	5	5	5
513 - 520	5	5	5	5	5	5	5	5
521 - 528	5	5	5	5	5	5	5	5
529 - 536	5	5	5	5	5	5	5	5
537 - 544	5	4	5	4	5	4	5	4
545 - 552	5	5	5	5	5	5	5	5
553 - 560	5	5	4	5	4	4	4	4
561 - 568	4	4	4	4	4	4	4	4
569 - 576	4	4	5	5	4	5	5	5
577 - 584	6	7	8	11	15	21	29	39
585 - 592	53	69	88	111	136	163	189	215
593 - 600	237	254	265	269	266	256	240	219
601 - 608	197	174	151	130	111	94	79	67
609 - 616	57	48	41	36	32	29	26	24
617 - 624	21	19	18	16	15	14	13	12
625 - 632	12	10	10	9	9	9	8	8
633 - 640	7	7	7	7	6	6	6	6
641 - 648	5	6	6	6	5	5	5	5
649 - 656	5	5	5	5	5	5	5	5
657 - 664	5	5	4	5	4	4	4	4
665 - 672	5	5	4	5	5	5	5	5
673 - 680	4	5	4	5	4	4	4	4
681 - 688	4	4	4	4	4	4	4	4
689 - 696	4	4	4	4	4	5	4	4
697 - 704	5	4	4	4	4	4	4	4
705 - 712	4	5	4	5	4	5	5	5
713 - 720	4	4	4	4	4	4	4	4
721 - 728	4	4	4	4	4	4	5	4
729 - 736	5	4	4	4	4	4	4	4
737 - 744	4	4	4	4	4	4	4	4
745 - 752	4	4	4	4	4	4	4	4
753 - 760	4	4	4	4	4	4	3	4

Table (2): HPLC raw data set

761 - 768	4	4	4	4	4	4	4	4
769 - 776	4	4	4	4	4	4	5	4
777 - 784	4	4	4	4	4	4	4	4
785 - 792	4	4	4	4	4	4	4	4
793 - 800	5	4	4	5	4	5	4	4
801 - 808	4	4	4	4	4	4	4	4
809 - 816	4	4	5	4	4	4	5	4
817 - 824	5	4	5	4	5	4	5	4
825 - 832	4	4	4	4	4	4	4	4
833 - 840	4	4	4	4	4	4	4	5
841 - 848	4	5	4	5	4	4	4	4
849 - 856	4	4	4	4	4	4	4	4
857 - 864	4	4	4	4	4	4	4	4
865 - 872	4	4	4	4	4	4	4	4
873 - 880	4	4	4	4	4	4	4	4
881 - 888	4	4	4	4	4	4	4	4
889 - 896	4	4	4	4	4	4	4	4
897 - 904	4	4	4	4	4	4	4	5
905 - 912	4	5	4	5	5	5	5	5
913 - 920	5	4	4	4	4	4	4	4
921 - 928	4	4	4	4	4	4	4	5
929 - 936	4	5	5	5	4	5	4	5
937 - 944	4	4	4	4	3	4	4	4
945 - 952	3	3	3	4	6	7	6	6
953 - 960	5	5	5	5	4	5	4	5
961 - 968	4	5	4	4	4	4	4	4
969 - 976	4	4	4	4	4	4	4	4
977 - 984	4	4	4	4	5	4	4	4
985 - 992	5	5	6	6	6	6	5	5
993 - 1000	5	4	4	5	5	4	0	0

Remarks concerning HPLC raw data set (Table 2) and GC raw data set (Table 1):

The numbers represent the raw data count number and the signal HEIGHT. Time data are coded in the consecutive number of the raw data count itself.

This will become clear immediately on consideration of the analog/digital conversion process controlled by a clock of "ppb precision" and repeated at constant intervals.

Let us set the conversion rate "R" at 10 data per second; that means: R = 0.1 s. Each consecutive conversion is done 100 ms after the previous one. The 346th height value corresponds to the time 346 * R in seconds. This allows the time to be encoded in the count number of the height data A/D step.

If a chromatogram is sliced up into peaks, one must know the starting time of the first peak point only, and the conversion rate R.

Table (3) is an example of a preliminary extract of raw data if the level and finer details of the baseline between peaks is unimportant.

Table (3)
A BASIC program found the start and end of peaks above a given minimum signal height and extracted all "peak raw data" from the total of the raw data set. All data belong to the raw data set of Table (1).

NR	SIGNAL	NR	SIGNAL	NR	SIGNAL	NR	SIGNAL	NR	SIGNAL	NR	SIGNAL	
338	16	339	15	340	15	341	15	342	19	343	44	
344	150	345	366	346	517	347	454	348	245	349	110	*****
350	43	351	24	352	19	353	17	354	17	355	16	

MAXIMUM = 517 AT T = NR 346 MINIMUM = 15 AT T = NR 339

NR	SIGNAL	NR	SIGNAL	NR	SIGNAL	NR	SIGNAL	NR	SIGNAL	NR	SIGNAL	
449	16	450	16	451	16	452	16	453	17	454	19	
455	32	456	68	457	157	458	286	459	379	460	395	
461	306	462	195	463	99	464	49	465	29	466	23	*****
467	19	468	18	469	17	470	16	471	16	472	16	

MAXIMUM = 395 AT T = NR 460 MINIMUM = 16 AT T = NR 449

NR	SIGNAL	NR	SIGNAL	NR	SIGNAL	NR	SIGNAL	NR	SIGNAL	NR	SIGNAL	
663	15	664	16	665	16	666	16	667	17	668	19	
669	25	670	39	671	67	672	109	673	177	674	254	
675	325	676	365	677	364	678	313	679	244	680	165	*****
681	104	682	68	683	43	684	30	685	23	686	20	*****
687	19	688	18	689	17	690	17	691	16	692	16	

MAXIMUM = 365 AT T = NR 676 MINIMUM = 15 AT T = NR 663

NR	SIGNAL	NR	SIGNAL	NR	SIGNAL	NR	SIGNAL	NR	SIGNAL	NR	SIGNAL	
1071	8	1072	8	1073	8	1074	8	1075	9	1076	10	
1077	12	1078	16	1079	22	1080	32	1081	46	1082	63	
1083	89	1084	122	1085	161	1086	197	1087	236	1088	271	
1089	293	1090	303	1091	296	1092	276	1093	242	1094	201	*****
1095	158	1096	122	1097	88	1098	62	1099	45	1100	32	*****
1101	23	1102	18	1103	14	1104	12	1105	10	1106	10	
1107	9	1108	9	1109	8	1110	8	1111	8	1112	8	

MAXIMUM = 303 AT T = NR 1090 MINIMUM = 8 AT T = NR 1071

NR	SIGNAL	NR	SIGNAL	NR	SIGNAL	NR	SIGNAL	NR	SIGNAL	NR	SIGNAL	
1848	8	1849	9	1850	9	1851	9	1852	10	1853	11	
1854	12	1855	14	1856	17	1857	20	1858	23	1859	27	
1860	33	1861	41	1862	49	1863	60	1864	72	1865	87	
1866	101	1867	119	1868	138	1869	153	1870	172	1871	191	*****
1872	208	1873	228	1874	248	1875	268	1876	284	1877	300	*****
1878	314	1879	323	1880	329	1881	328	1882	323	1883	313	*****
1884	297	1885	275	1886	251	1887	222	1888	192	1889	166	*****
1890	138	1891	112	1892	93	1893	73	1894	57	1895	45	*****
1896	37	1897	29	1898	23	1899	19	1900	16	1901	14	
1902	12	1903	11	1904	10	1905	9	1906	9	1907	9	

MAXIMUM = 329 AT T = NR 1880 MINIMUM = 8 AT T = NR 1848

On comparing the raw data of peak 1 with those of peak 5 in Table (3), one immediately realizes the weakness of constant rate A-to-D conversion: the number of available measurements may be too small at the beginning and too large at the end of the chromatogram if isothermal GC is used. In the most widely employed temperature-programmed mode of GC, however, constant rate A-D conversion is the right procedure.

Table (4) now extracts the maximum time values for each of 6 homologues (*n*-alkanes) including 3 values before and 3 after each peak maximum.

Table (4)
A further extract now from the data in Table (2): Only the seven points around the peak maxima of the raw data are found by a simple BASIC program. This material is ready to be used for retention time optimization.

NR	SIG	NR	SIG	NR	SIG	NR	SIG	NR	SIG	NR	SIG	NR	SIG	
343	44	344	150	345	366	346	517	347	454	348	245	349	110	*1
457	157	458	286	459	379	460	395	461	306	462	195	463	99	*2
673	177	674	254	675	325	676	365	677	364	678	313	679	244	*3
1087	236	1088	271	1089	293	1090	303	1091	296	1092	276	1093	242	*4
1877	300	1878	314	1879	323	1880	329	1881	328	1882	323	1883	313	*5

SUMMARY :

PEAK NR	MAXIMUM	AT I=	(IN SEC.T=)
1	517	346	34.6
2	395	460	46
3	365	676	67.6
4	303	1090	109
5	329	1880	188

In order to extract these more important 43 raw data from the total of 2000 in our data example we just used the following simple algorithm in BASIC:

```
100 SS = 1 : REM (slope sensitivity)
110 MM = 100 : REM (minimum signal height which is
accepted as value for a peak raw maximum)

120 FOR I = 1 TO DA
130 M = D%(I+1)
140 IF M > D%(I)+SS AND D%(I+2)<M OR
D%(I+3)<M OR D%(I+4)<M AND M>MM THEN PZ
= PZ+1: MX(PZ)=M : TX(PZ)= I+1
150 NEXT
160 PRINT ..... (all the results)
```

DA was the amount of raw data
D%(I) the Ith raw height signal
PZ is the peak found counter
MX(PZ) the PZth peak maximum
TX(PZ) * DT the corresponding time to the PZth peak and DT the A/D conversion rate.

4.2 Practical versus Theoretical Peak Maximum

Which model will allow us to properly INTERPOLATE time data around the top of a peak in order to find the best possible peak MAXIMUM under noise conditions as well as with flat peak top data?

Most of the publications treating retention time calculation based upon raw data around a peak top use 2nd order regression. Let us compare 2nd, 3rd, 4th, and 5th order polynomial regression AND interpolation of peak tip data. We use real-life data in order to avoid misinterpretation. If we were to use synthetic Gaussian peaks, we would end up with useless results.

Real peaks are never Gaussian in shape, although the opposite is stated in some of the latest publications. The worst procedure is to SIMULATE GC by sine wave or Gaussian peak generators and to sell conclusions based upon this experimental nonsense.

The peak shape at top of the peak may be as noisy as the whole raw data material. Spikes would falsify the retention time value. Noisy data could be treated by LEAST-SQUARES techniques and so the 7 data points (including 3 on the left, and 3 on the right of the raw peak maximum) could be interpolated by 2nd, 3rd, 4th, and 5th order least-squares polynomial regression.

If spikes are absent and the raw data are NOT noisy—this is established by analyzing baseline data—then a much better way of arriving at retention time data is raw data interpolation by 2nd to 5th order polynomial INTERPOLATION, or by using cubic spline techniques. We have named "least-squares polynomial interpolation" polynomial regression to distinguish it from real point-to-point (soft) interpolation.

The difference between the two techniques has already been mentioned and can be seen in Figure (5); however, additional data should be treated and the result analyzed to make clear which technique is useful under what data conditions.

Graphics data display in connection with mathematical model treatment is a great help in understanding the problem.

The following Table (5) compares the theoretical maximum and the statistical data found when 2nd, 3rd, 4th, and 5th order LEAST-SQUARES polynomial regression has been used with the "seven peak top raw data" of Table (4).

Table (5)

Peak no.	1	2	3	4	5
raw max.	346	460	676	1090	1880
2nd order	346.2	459.7	676.4	1090.1	1880.5
corr.coeff	0.949	0.956	0.987	0.998	0.998
3rd order	346.4	459.3	676.5	1090.2	1880.5
corr.coeff	0.955	0.981	0.993	0.999	0.999
4th order	346.2	459.6	676.4	1090.2	1880.5
corr.coeff	0.999	0.999	0.999	0.999	0.999
5th order	346.2	459.6	676.4	1090.1	1880.5
corr.coeff	0.999	0.999	0.999	0.999	0.999

(corr.coeff = correlation coefficient; cubic spl = cubic spline interpolation technique.)

Now we will treat the whole material by SECOND ORDER, FIFTH ORDER direct polynomial interpolation. Each raw data point is treated as correctly positioned and 2nd order and 5th order polynomial curves, respectively, connect the raw data points. Cubic spline techniques are also used to connect each raw data point.

On the basis of these connecting curves the position of the theoretical peak maximum is calculated and listed in the table. Now we have no statistical data for a quality decision regarding the optima found. Graphics display is fine for fundamental studies, but in practice there is no time to look into those details under routine laboratory stress.

Here CROSS CORRELATION of data will help, it will again be discussed by comparing the "theoretical" retention value obtained in Tables (5) and (6)

Table (6)
Dead time calculation by "iteration" and "constant polarity procedure" 3, 4, [6].

Peak no.	1	2	3	4	5
raw max	346	460	676	1090	1880
2nd.int.	346.2	459.7	676.5	1090.1	1880.3
5th.int.	346.3	459.7	676.5	1090.1	1880.3
cubic spl.	346.3	459.7	676.5	1090.1	1880.3

Note: Here we have to add an important note about polynomial least-squares interpolation (regression):

The data given in the Table have been calculated by entering the raw data NUMBER (1...7) as the X-axis values, and the signal height as the Y-axis values. Otherwise 5th order polynomial regression would have been impossible *e.g.* the data corresponding to peak 5.

This can be seen on comparing the following data set:

Table (7)

X-data	1	2	3	4	5	6	7
Y-data	300	314	323	329	328	323	313
calc.	300	313.8	323.4	328.5	328.4	322.8	313

The data show a defined maximum.
Regression quality (mean of diff. between given and calculated Y-data): 0.07 ± 0.05 %
Determination coeff.: .99902
Data see Figure (5).

Now using the raw data count number as X-axis value

	1877	1878	1879	1880	1881	1882	1883
Y-data	300	314	323	329	328	323	313
calc.	312	314.2	316.3	318.5	320.7	323	325.3

Regression quality 2.2 ± 1.66%.
Determination coeff. 0.2117.
No defined (usable) maximum. See Figure (9).

In the practice of automatic polynomial data regression there is no problem associated with the use of this legal data transformation, as the peak raw signal counter, together with the maximum-finding algorithm, can find the data 1...7 around the peak top. The interpolated retention time can easily be refound, as the computer remembers the "address" (consecutive count number) of the raw maximum. If this were "1880" as in our example, then "X = 1" would correspond to 1880 − 3, X = 4 would correspond to 1880, X = 7 would correspond to 1880 + 3, and a found OPTIMUM of "3.5" would lead to 1880.5 as the interpolated final RAW retention time. Multiplication by the rate of A/D conversions leads to the final optimized retention time.

4.3 Which Retention Time Data Are Correct?

Now we must use cross correlation.
Answering this question will need some computer time, but supplies us with a statistical quality number. This number expresses the "quality" of the dead time. In addition, it supplies us with information about the possible presence of false time data. Or of data which have been systematically altered by experimental errors (temperature, pressure, nontightness) or by sample influence (overloading, polarity change by solvent effects, *etc.*).

It is incredible how often retention time data of single peaks are falsified. As this dramatically influences the identification and also the quantitative correction factors, we are grateful for any information forthcoming from "CROSS CORRELATION" to alarm us in cases of wrong data. We will find out which of the values fit best into the chromatography law of linear correlation between log(adjusted retention time) and molecular weight. Instead of using the molecular weight as such, we use the carbon number or the "member number" within the homologous series. In gas chromatography, this member number can best be expressed by the retention index in the case of alkane homologs.

As we have used the test mixture hexane...decane in undecane, all data are available for cross correlation. We will let the computer experimentally deduct an approximative value for the dead time, then check the linearity of the correlation log(tms − approx. dead time)

over retention index. Then the "quality number" of linearity, which is the square of the regression coefficient, will be used to tell the computer which improvements of the dead time approximation it has to make. The computer will iterate, until the linearity of log(tms − tm) over index can no longer be improved.

We call this procedure the "iterative dead time calculation mode."

Note: Some chromatographers misunderstood this procedure and thought that it falsifies the dead time. They consider that dead time value to be of importance which is the natural retention time of a compound of partition coefficient K = zero (some hypothetical substance which may not actually exist). We find another dead time value to be of importance: *i.e.* the value which is a precise measure of mobile phase flow (for automatic system optimization on-line). This is the value which allows most accurate identifications.

Wrong non-adjusted retention time data WILL REMAIN WRONG even after correction with the calculated dead time. They will lead to wrong adjusted retention time data. But we can easily see whether anything is wrong just by inspecting the "cross correlation quality numbers".

Up to now, our discussion of raw data has been non-specific with respect to the chromatographic technique. Everything said was just as true for high pressure (or high performance) liquid chromatography as for gas chromatography and in the latter case for GC with packed columns just as well as GC with capillaries.

Difference might also exist with respect to absolute values: we may be dealing with very short packed columns in GC or very high pressure in HPLC on the one hand, and very long packed columns in GC or very badly installed HPLC columns of very low resolution efficiency on the other.

We have not yet discussed the special conditions prevailing with 2m long 0.05 mm i.d. super high speed capillaries; however, A/D converters are available to handle this speciality. The only problem is that the fast detector amplifiers necessary are not yet available, or affordable for routine work. And the star operator who handles the sample injection problem is not yet available.

So let us discuss normal ROUTINE with data of comparatively speedy chromatography in the 1 to 5 minute range.

In our subsequent discussions we must distinguish between dead time calculations in GC and HPLC.

Let us consider GC first:

4.4 Dead Time Calculation in Gas Chromatography

At least four algorithms are in use to calculate dead time data, authored respectively by:

1. *Peterson/Hirsch* [3] = Procedure 1
2. *Said* [4] = Procedure 2
3. *Kaiser/Sevcik* [5] = Procedure 3
4. *Kaiser/Ebel* [6] = Procedure 4

4.4.1 Principles of Dead Time Calculation Procedures

Procedure 1:

tA, tB, tC = non-adjusted retention times of three consecutive homologues

$$tm = \frac{tA * tC - tB^2}{tA + tC - 2 * tC}$$

Problem: no information in case of falsified data. No quality index available.

Procedure 2:

See pages 148 to 150 in [4] = *Abdel Salam Said,* "Theory and Mathematics of Chromatography", a volume of the green series on CHROMATOGRAPHIC METHODS (1981).

Procedure 3:

Basis is the fact that the "selectivity" of a stationary phase is constant throughout a series of homologues. This is more or less true with polar substances and not true for the first members of the homologous series. But it is perfectly true in GC for *n*-alkanes on semi-polar phases.

Sevcik in his paper [5] did not treat "tm calculation" based upon his results, but *R.E.K.* used the results of his paper to write a BASIC program, which allows very fast calculation of tm data AND offers statistical data for cross correlation of the raw data "quality".

Procedure 4:

We let a computer deduct guessed time values and systematically check the resulting linearity of log(tms − "guessed tm") over the carbon number. We start with 0.9*tms of the smallest tms value given in the raw data set. This is a time consuming iteration procedure. *Ebel* [6] improved our first program to make it fast. The Procedure 4 offers the best value for tm and statistics data, which allow for cross correlation and raw data check.

In the following, only those procedures are used to calculate the dead time tm which allow additional tests with the tms values. Procedures 1 and 2 work with any data set, including those without any practical significance. Any three log data can be linearly correlated. This is the basis of Procedures 1 and 2.

In the case of Procedures 3 and 4, examples of data are given, the algorithm listed, and the programs cited.

Procedure 1 can be used with 3 consecutive test homologues only; in practice it is better to use more than four normal alkanes. Example: *n*-hexane, *n*-heptane, *n*-octane or *n*-C12, *n*-C14, *n*-C16, *n*-C18 or methane, ethane, propane, butane.

The concept underlying the algorithm of Procedure 3 is that the relative adjusted retention of consecutive pairs of test compounds have equal "polarity" and that underlying the algorithms of Procedure 4 is a linear relationship between log(tms − tm) and the homologue number

Procedure 3 can be used only with data for consecutive homologues. At least four tms values are necessary; *n*-alkanes are preferable, but higher members of the homologous series of fluorinated esters, halocarbons, *etc.*(ECD !) or just esters may be used in order to get acceptably "good data" for a wide variety of stationary phases.

What is meant by "good data" will be discussed in the Section on cross correlation of retention time values. Procedure 4 finally is the most flexible one and can be used with any group of at least four substance of KNOWN RETENTION INDEX.

Cross correlation of retention time values with retention index data, determination coefficients, tm standard deviations, and tm differences will help us to choose the best possible CORRECTED retention time values.

We use the data from Table (4), page 58, and Table (5), page 60, respectively, which have been compiled from the values of the alkane test run already treated.

tms	C 6	C 7	C 8	C 9	C 10	Result CC
raw data	346	460	676	1090	1880	tm = 222.80 ± .58% (B = .9999963) tm = 220.40 (Pr. 4) (B = .999999)
interp. 2	346.2	459.7	676.5	1090.1	1880.3	tm = 223.91 ± .062% (B = .9999934)
interp. 5	346.3	459.7	676.5	1090.1	1880.3	tm = 224.30 ± .1% (B = .9999924)
regr. 5	346.4	459.6	676.5	1090.1	1880.5	tm = 224.74 ± .242 (B = .999993) tm = 221.69 (Pr. 4) (B = .999999)

"Pr.4" means: Procedure 4 (list above) has been used to calculate tm and to check the quality of the data by the square of the regression coefficient.

Note: Although we used 0.1 s precision "optimized" retention time data, the results obtained for "tm" are not exciting!

However the discrepancies with "good data" are below 1% rel, when comparing differing techniques of calculation. The most accurate tm value is the one which gives the largest coefficient of determination B.

Two differing dead time calculation procedures have been used. Besides furnishing a value of the dead time, they also supply statistical data, *viz.* the standard deviation of tm and the coefficient of determination of the correlation between log(ts) and the retention index.

The reader may perhaps not yet see what cross correlation of raw retention time data with retention index and GC laws has to offer in this special case, because the raw data used in our example (selected by statistics from many hundred data sets) are too good.

In order to visualize the potential of this approach, let us assume that our raw data have been measured to within the full second only, instead of with 0.1 second precision, and then "corrected" by interpolation to the 10 millisecond level.

Let us just change our raw data set towards this low precision level, although the digital numbers are not accurately SECONDS (see below). If the measurements on our alkane test mixture had been performed at a 1 second precision level we would have obtained the following raw data:

C 6	C 7	C 8	C 9	C 10	tm, ± standard dev. (B, determin. coeff)
340	460	670	1090	1880	tm = 204 ± 7.0% (B = .99987) ! ! tm = 213.86 (Pr. 4) (B = .999905)

The reader might complain that the data system should have been at least sufficiently qualified as to round up to the full second. Some do this, and in that case the data would have read

350	460	680	1090	1880	tm = 234.24 ± 3.42% (B = .999935) tm = 227.8 (Pr. 4) (B = .999973)

Another objection might follow: One could state that it is IMPOSSIBLE to start a chromatographic analysis more precisely than to within one second. The manipulation involved in injection is time consuming. The dissolution step (to dissolve the sample in the mobile phase) is time consuming. The mobile phase flow "shock" needs time *etc., etc., etc.* All this is quite true but does not cause any serious problem. Procedure 4 for dead time calculation is not burdened by any of the practical limitations of the type just mentioned. The resulting tm value is free from systematic start time errors—the procedure CORRECTS for systematic start errors. However: the value found will be useful for additional (cross correlated) chromatography data only if the final result is as correct as is technically possible.

For an acceptable result we need raw retention time data precise to the 10 millisecond level, and accurate to at least tenths of a second. However, any systematic error of too early or too late a chromatogram start are corrected automatically with respect to the "chromatography START position" in the chromatogram. This might become clearer on considering the following remark: Although the technique of chromatographic separation starts with the moment of injection, chromatographic SEPARATION *per se* starts only AFTER the dead time is over. We call this point in the chromatogram the position "k = 0" and will explain this more in the Section "System Analysis".

Some pertinent data illustrating that systematically wrong start time data do NOT influence the correct data for the "k = 0" point in elution chromatography are given in the Section "Dead Time Calculation in HPLC". Procedure 4 for dead time calculation holds true for GC as well as for HPLC.

4.5 Analysis of the tm Data

Value	Procedure	Quality number	
222.80	3 with raw data	± 0.58%	B = 0.999993
220.4	4 with raw data		B = 0.9999992
223.90	3 with interp.2nd	± 0.06%	B = 0.9999934
220.9	4 with interp.2nd		B = 0.99999999
224.3	3 with interp.5th	± 0.1%	B = 0.999992
224.73	3 with regr.5th	± 0.24%	B = 0.999993
221.69	4 with regr.5th		B = 0.9999999

If one checks which raw data errors exert the greatest detrimental effect on the "quality numbers", then the main blame lies with the raw data of the early eluting peaks. Systematic errors arising in sloppy integrators are responsible for systematically wrong raw data sets. The culprit may be an electronic integrator of a class which enjoys great popularity on account of their (too) large time constants. This effect could be misleading. One might even welcome the visible effects of such "data massage". Printer plotters plot much nicer baseline when fed with highly smoothed signals; however, the true peak shape has gone, and not only the peak shape. The retention time has been falsified! And so has everything else that depends upon the retention time. And that is just about everything in qualitative chromatography...

So we tried changing the RAW retention time of the first peak from 346.4 to 346.1 seconds, and the value for peak two from 459.6 to 459.7 seconds.

Results

tm			
223.59	3 math.exp.	±0.02%	B = 0.99999
220.81	4		B = 0.99999999

Both dead time calculation techniques afford high quality checking data for the chromatography law (Procedure 3: the polarity of homologous pairs remains constant throughout the series; Procedure 4: log(tms − tm) is linearly correlated with the molecular weight). The two results differ by more than one % rel.

Which value is the BETTER one? That tm value which shows the most perfect linear regression line, because it is INTERPOLATION which we use in all qualitative work.

Data analysis in this chapter ends up with the result: The dead time value corresponding to the raw data IMPROVED by fifth order polynomial regression is

tm = 220.8 s (or time units)

Compare this tm = 221 with tms = 346 of the first peak!

Deduction of tm led to the adjusted retention time of ts = 125 s (or time units) and a + 1 % systematic tm error led to a 2% systematic retention time error for the first peak (at k = 0.5)

If we had accepted the full second (or time unit) data with tm = 204 or 234 s (or units) we would have ended up with a ts of 25% uncertainty in the not too early range of a chromatogram.

Let us close this section by repeating:

TIME DATA IN CHROMATOGRAPHY CANNOT BE PRECISE ENOUGH.

If chromatography is done on the several minutes' elution time scale, then time precision MUST be on the 10 millisecond level. This is especially important in case of fast capillary GC.

The tool we used to produce the data set in the above example was a 15 m fused silica capillary. The sample we used was hexane to decane, the temperature around 100 °C, the absolute time data will be seen in section SYSTEM ANALYSIS.

It was necessary to use several thousand words to explain the background of the story of time interpolation. But for calculation of the tm value and its statistics (from 4 single data resulting in the standard deviation) and the log(ts)/C number linearity regression test (resulting in the determination coefficient) for the data set used a simple computer program (compiled BASIC) would need the following computer time (Apple II)

Reading the data:	<2 s
Calculating tm + s s(tm):	<1 s
log regression	<1.5 s

tm is therefore really a measure of flow for

method check
system analysis (see later)
optimization

but needs very precise raw data.

And tm is needed for calculating the RETENTION INDEX in isothermal GC and the computer needs the retention index as THE ONLY qualified ADDRESS of a substance-specific correction factor in automatic but critically controlled quantitative analysis. This matter will be discussed in a later volume.

The discrepancy in tm data of Procedures 3 and 4 warrants careful thought!

A higher precision level, or a lower detectability level, or better accuracy has always been the starting point for improving a method, detecting an important detail, OR CHANGING A THEORY!

Further arguments relevant to this point will follow in the Chapter entitled "Shape Analysis". First of all, though, we have to discuss

4.6 Dead Time Calculation in HPLC

This Section can be kept very short:

Many chromatographers have doubts about the existence of any dead time in HPLC. They then declare that one cannot calculate something which does not exist... Yet many chromatographers use "k values" in HPLC which are

$$(tms - tm) / tm$$

which brings us back to familiar ground.

Within homologous series of compounds, a relatively good linear relationship exists between the log of ts and the molecular weight of the homologues.

There is—when using "similar" compounds—quite a good linear correlation between the non-adjusted retention time and the partition coefficient K. Thus there is quite a good correlation between the partition coefficient K and k = (tms − tm)/tm.

We therefore tested, in thousands of HPLC experiments, the correlation between log(ts) and the molecular weight and found adequate but never PERFECT linearity between log(ts) and the carbon number of the homologues. There follow some arguments which should stimulate our interest in a fast measure of the mobile phase flow in HPLC to permit autooptimization:

● HPLC has a great appetite for mobile phase (a classical packed 4 mm i.d. column needs about 100 to 200 l mobile phase per month when running round the clock). A liter of pure water may cost 10 to 20 dollars. Should we not do everything in our power to OPTIMIZE such an expensive technique?

tm would provide an automatic measure of the mobile phase flow; in fact, the correlation between 1/tm and flow is perfect with a linearity determination coefficient of 0.9999, when flow measurements are done precisely and accurately, and tm data are calculated from retention time data of 0.1 to 0.01 second precision

Apart from the rather primitive formula:

dead time = column dead volume / flow

an on-line tm algorithm has yet to be found in HPLC. So we used our "log(tms − tm) to homologous carbon number"—iteration procedure. This was described in detail in the Section "Dead Time in GC".

We have not yet found a test mixture, with which the Procedure 3 (constant value of polarity in a homologous series) holds true. Unfortunately, therefore, only one procedure is available. The experimental details are described in "Optimization in HPLC" [7] by *Kaiser* and *Oelrich*.

A table of results should be suffice as an invitation for the reader to test the procedure in practice.

Using the "tm iteration program algorithms" at the end of this volume—available as complete working program on disk from Institute for Chromatography—the reader can easily write his or her own program. Since iteration processes are time critical, we therefore preferred GOSUBS over FOR NEXT loops.

One calculation done with a compiled BASIC version of the iteration program needs about 20 seconds for 5 retention time raw data.

A complete HPLC raw data set is given in Table (2), page 69, for colleagues who want to examine the details.

We used a 115 mm long, 4 mm i.d. steel column packed with 3 micron spherical reversed phase material type ODS 2 (octadecylsilane, chemically bonded).

The column was packed by *K. Bischoff*—Stuttgart [8] who routinely packs 3 micron columns with more than 10,000 plates per 100 mm column length—see SYSTEM ANALYSIS HPLC. We ran the analysis in a home made HPLC instrument with a Spectra Physics pump model 8750. We used a UV detector at 254 nm and fed the detector signal directly into the 12 bit IMI Adalab A/D-D/A converter board (in an Apple II). We converted the raw signal at a rate of 0.2 to 0.6 second. 1000 (1024) raw data had been collected, representing a HPLC test run of 300 to 600 seconds total elution time. As a test sample, we used the homologous series of 3,5-dinitrobenzoic acid esters C3...C7.

The chromatogram showed impurities which were poorly resolved and which eluted prior to the first main peak, as well as a negative signal, which was the sample solvent. The mobile phase was acetonitrile/water 85/15 v/v.

Now the peak maxima have been found and analyzed in a range of 7 raw data (3 left, 3 right of the maximum) by 2nd and 5th order polynomial regression.

As the HPLC data were much "softer" than the GC capillary data due to a much broader peak width at half height—see the Section entitled "Shape analysis"—the polynomial regression technique could fit much better. Here are the data:

Non-adjusted retention time interpolation:

Peak no.	tms opt 2nd	tms opt 5th order polyn. regr.
1	236.9	236.7
	B = 0.973	B = 0.99999
2	286.2	286.1
	B = 0.994	B = 0.99999
3	354.6	354.4
	B = 0.986	B = 0.998
4	452.9	452.9
	B = 0.996	B = 0.99999
5	593.6	593.4
	B = 0.978	B = 0.9998

Conclusion: The fifth order polynomial regression of the 7 raw data around the peak top fits so much better than the second order polynomial regression that the fifth order procedure is strictly recommended.

There are distinct differences in the tms optima obtained by the two interpolation techniques. We will examine their influence on dead time calculations.

From the non-adjusted retention time data we proceed to the dead time calculation experiment:

We used Procedure 3 as described above—it failed. The computer program protested against the data and rejected the input, complaining that there is NO linearity between log(tms − tm) (Procedure 3) and the carbon number of the homologues, see Figure (9).

The data are unacceptable:

tm	Standard deviation	Determin.coeff.
160.6	± 2.1%	B = 0.997

and with the fifth degree regression tms data:

160.0	± 2.2%	B = 0.998

We then tried the iteration Procedure 4, which works provided that there is a linear correlation between the adjusted retention time and the homologue carbon number.

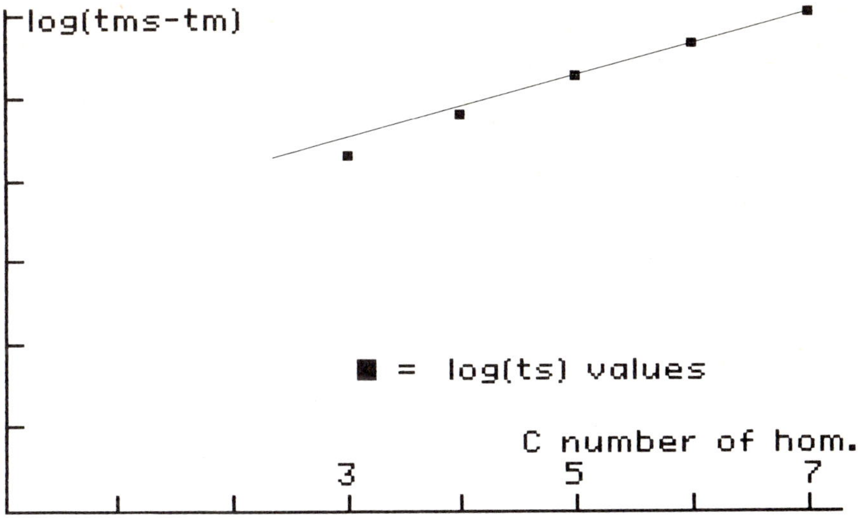

Figure (9)
The log(adjusted retention time) is here NOT linearly correlated with the carbon number of homologues.

Although not particularly good, these data are at least acceptable:

Results:

Based on 2nd order regression tms values:

tm = 121.3 ± 0.2 B = 0.99998

Based on 5th order regression tms values:

tm = 120.9 ± 0.2 B = 0.99998

Closer examination of the raw data table might reveal the reappearance of the "negative peak" at raw data numbers 100 to 120.

The tm value of 121 units is in no way nonsensical. We had not discussed what happens with the data if the chromatographer starts the run too late or too early. Let us use the, by no means excellent, HPLC dead time calculation to check what happens with falsified values, especially with a wrong starting point. Let us falsify the data by minus 100 units (100 time units too late!). The results based upon the fifth order polynomial regression data are:

tm (started 100 units too late)
= 20.1 ± 0.13 B = 0.99998

Add the 100 missed time units and tm = 120.1

Let us now consider the case of much too early a start:

tm (started 100 units too early)
= 220.9 ± 0.33 B = 0.99998

ERRORS in the starting time of the run do NOT falsify the value of k = 0 as the starting point of chromatography in the chromatogram. In this way the "dead time" is truly eliminated. This is just as true with positive as with negative errors with respect to the wrong starting times.

We now test the influence of the type of data we use as "homologue value" (carbon number times 100, homologue number, true molecular weight, or whatever):

Instead of 300 — 400 — 500 — 600 — 700 as in the case of the homologue number, we used data multiplied by 1000 each.

Results:

tm = 120.7 ± 0.2 B = 0.99998

The procedure is safe.

Run a real isocratic run of a TEST series of compounds, which belong strictly to a homologous series (do NOT use benzene, naphthalene, anthracene, naphthacene). It is permissible to use propyl, butyl,...nonylbenzene, or series of esters (but no compounds having too strongly solvated an environment around the molecule) and you will get an acceptably correct dead time in HPLC, VALID for THOSE compounds only which have a molecular shape SIMILAR to that of the members of the homologue series.

There is in fact no uniform tm value in HPLC; in other words the dead time is the value of available volume of mobile phase for the molecules of interest divided by the mobile phase flow. "Available volume" means space in which molecules can move.

Example: let us assume that we are dealing with a superfine pore stationary phase, where a considerable part of the pore volume is enclosed by very narrow channels. Let us further assume that the molecules to be chromatographed are too large in diameter to enter the pore channels. This is the case, for instance, in reversed phase chromatography in water/methanol mobile phases for various free organic acids. Barbituric acid is an example of a molecule surrounded by a huge cloud of water molecules in relatively strong association.

Aquo-barbituric acid molecules have a volume too large for some of the mini-pores in the stationary phase, which nevertheless are filled with mobile phase. Thus the mobile phase is there, but the aquo-barbituric acid cannot move in. Hence, this part of the mobile phase is NOT available for aquo-barbituric acid. Therefore, in accord with the law

tm = dead (unavailable) mobile phase volume / flow

tm is large for small molecules and small for large molecules.

Therefore aquo-complexed barbituric acid can easily elute with SHORTER times than the normal—non-complexed—barbituric acid esters. Barbituric acid is excluded from a part of the mobile phase volume. We thus have a case of EXCLUSION chromatography. This is the origin of the problem of differing dead time data. tm depends on the "available" mobile phase volume and it can be calculated according to Procedure 4 only by using PROPER homologous series.

However, we still need tm data, as the chromatographic address

k = (tms − tm)/tm

is the only qualitative measure we have in HPLC and can use for further evaluations.

Separation of substances which are truly soluble within the whole pore volume takes place within the range

k = zero to k = (stop elution time)/tm

The "stop elution time" is the time at which mobile phase flow is stopped and back flushing started (hopefully backflushing is indeed started, but this tends to overstep the borders of computer chromatography).

At k = zero the separation efficiency is zero and a huge number of substances can be eluted together unseparated. To know more about this important boundary value in elution chromatography we should start SYSTEM ANALYSIS. This is not some kind of game of numbers with no practical implications for chromatography; it has now become an easy task by using computers.

System analysis does not necessarily have to be treated differently for GC and HPLC, as language, principles, algorithms, and programs are the same in both fields.

5 System Analysis in GC and HPLC

Principles, possibilities, limitations, and results of peak shape analysis in gas and high pressure liquid chromatography are discussed. The two raw data sets used in all chapters up to now are the basis for all calculations.

Peaks in GC and HPLC are analyzed for their Gaussian—or better NON-Gaussian shape, as this allows choice of the procedures to be used when peak width data are to be calculated.

It is not permitted to use anything other than just the real peak width at half height "measurement". As the most readily available discrete data are most often those which need to be smoothed etc., statistical methods are necessary to get an acceptably accurate and precise peak width measurement. All values describing SEPARATION are derived from accurate retention time and sufficiently precise peak width data.

It is shown that for inter-and extrapolation, as well as for optimization, a LINEAR regression model is beneficial, with significant implications for the ABT concept one of us developed a decade ago.

Concentrate

Basis of system analysis—how good is the column, where is its physically practical separation optimum?—are correct and precise enough peak width data and acceptable mathematical models. Peaks are non-Gaussian and their time data—peak width at half height—cannot be precise and accurate enough. The usable procedures and results of critical comparison are discussed in depth.

5.1 Peak Shape Analysis

The only directly measurable information we have is retention time, some knowledge about signal height, and the shape of peaks. That's all that chromatography has to offer.

Peak width data are of importance for assessing the quality of a separation as well as the quality of a column. But peak width and peak shape data reflect a combination of **instrumental** contributions, **substance** contributions, and **user** effects (how injection is performed, how the sample is transferred from its highly concentrated state into the mobile phase). We have mini and macro mobile phase dead volume effects, or volumes filled with mobile phase when back-mixing occurs. We can extract a considerable amount of information by critical, intelligent analysis of all details of the peak shape. Because this is the ONLY record we have of the combined effects of the instrument/sample/user, it is imperative that we understand the language in which it is expressed.

Such an understanding is best acquired by again examining our raw data with the intention of taking a critical look at the peak shape.

We proceed by:

a) Plotting the raw data on a computer graphics display;

b) finding the peak maximum by interpolation (as discussed in the Section above);

c) finding the peak width at half height;

d) calculating the standard deviation of the peak and comparing it with a theoretical standard deviation the peak MUST have if of symmetrical Gaussian shape;

e) calculating anything else of interest, for instance the center of gravity.

The following illustrations—three from gas chromatography data and three from liquid chromatography data, with early (fast) and late (slow eluting) peaks—will clarify what is meant by SHAPE analysis. We use the raw data already given in the two long tables on pages 41—46 and 54—56.

Peak width at half height, and at the height levels of interest in classical models of peak shape in chromatography = Gaussian at 88%, 60.7%, *etc.* of the peak height, at 15% and 85% for asymmetry qualification data are of **not only** theoretical interest.

Asymmetry data can furnish information about mechanical errors in chromatographic systems and can also help in proper identification. They can offer a guide for improving the stationary phase chemistry or improving the mobile phase composition. They can afford additional data about temperature homogeneity or the sampling process. Above all, however, peak width data hold the key to mechanical optimization of a chromatographic system and are eminently useful for showing the manager how good one's own column's are, *etc.*

However, peak width data are not within easy reach of the user: it is most time consuming to measure them and it is almost hopeless to want to measure them as well as necessary. The reader may already have developed an aversion to the phrase "not precise enough", but such is real-life chromatography:
"peak width data cannot be measured precisely enough" is so very true.

It is especially true with high resolution columns. Very soon one produces inflationary numbers of no practical meaning when using the peak width data as such in order to calculate plate numbers or plate height values.

5.2 Shape Analysis Procedure

As described earlier we will

- interpolate the optimized retention time
- find the baseline
- interpolate the peak shape at the top of the peak by 5th degree polynomial regression
- deduct the baseline from all peak signals
- find the half height
- use two to four raw data around the half height to the LEFT and to the RIGHT of the peak sides for linear regression
- cut these two lines with the parallel to the baseline (or X axis) through the peak half height and thus arrive at the first value of importance:

  b0.5 (s)

We will measure or calculate and express this and all other peak shape data in seconds (as far as they belong to the time axis) and in volts or amperes for the signal (Y axis).

But first of all we shall look at the raw data. We shall **cut** the **peaks** out of the chromatogram and store the start address (height and time counter number of the digitized signal) and the end address together with the enclosed signal data in a raw data file. Data files soon become numerous. We therefore adopt a „naming convention": each section representing a peak CUT out of a chromatogram raw data file is named as follows: »(name of raw data file). PK(number)«

Example:

If the raw data set is named

L16-233.1009/.12/.55

(which means: liquid column chromatogram
instrument no. 16
done on day no.233 of the year
1009th minute of the day with
500 mV full range
for 12 bit resolution
conversion rate 0.55 seconds

then the raw data of PEAK one, two, three, n of this data set is named
⟫ L16-233.1009/.12/.55.PK1
2
3....

This naming convention might appear to be overdone. But the reader should think about automation of data analysis in the lab! It would be a disaster to enter the computer era of laboratory work only to find a far greater workload than before the advent of computer help!

Here is a little BASIC program which has the computer do the job:

```
100 FOR I = 1 TO N
200—230 OPEN....READ....INPUT NAME$".PK";I
300 (DO THIS AND THAT WITH THE DATA)
400 CLOSE
500 NEXT
```

Let us examine graphics printouts of the capillary GC raw data set and the HPLC data set used in this Section up to here.

The Figure shows

- The position of the retention time—ticks are only to

inform about the position and the total of the raw date per peak.

- The value of standard deviation s.
- The position of the peak perpendicular.
- The peak width at 50% of interpolated height.
- The center of gravity.

One look will tell us everything about symmetry and deviation from Gaussian shape. A symmetrical Gaussian peak—according to *Golay* [9] such a peak is theoretically possible only on injection of zero grams of substance into an infinitely long column and avoiding use of a detector (because its volume would again disturb the Gaussian shape)—simply does not exist in practice. Use of the peak width at 60.7% height as the value of twice sigma (standard deviation) is therefore highly questionable. Apart from this very serious objection, use of the peak width at half height is preferable for another reason:

- Anybody who continues to believe in Gaussian shape could thus measure "s = sigma" more simply:

s = b0.5 * 1/ sqr(8 * ln (2))

Note: For any width b at relative height q:

s = b / sqr (8 * ln (1/q))

Here are the results:
GC peak one...five: see Figures (10)...(14)

HPLC peak one...five: see Figures (15)...(19).

Critical scrutiny of this series of peak shapes will show:

1) All peaks are non-Gaussian; however, the capillary GC peaks deviate not too far from a symmetrical signal.

Note: The standard deviation sigma stands for an unlimited amount of "data"—in this case the number of molecules producing the analog signal. This data "density" is not given when digitization takes place. Therefore "s" is the approximate value to be used instead of sigma.

2) The HPLC peaks are highly non-symmetrical, as is clearly demonstrated by the distinct difference in the

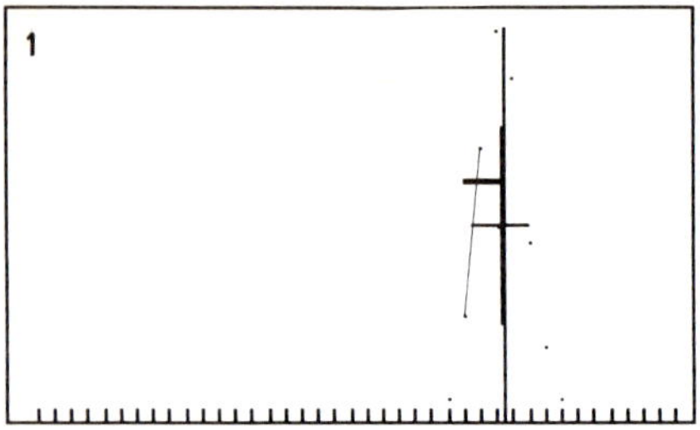

Figure 10 (peak 1)

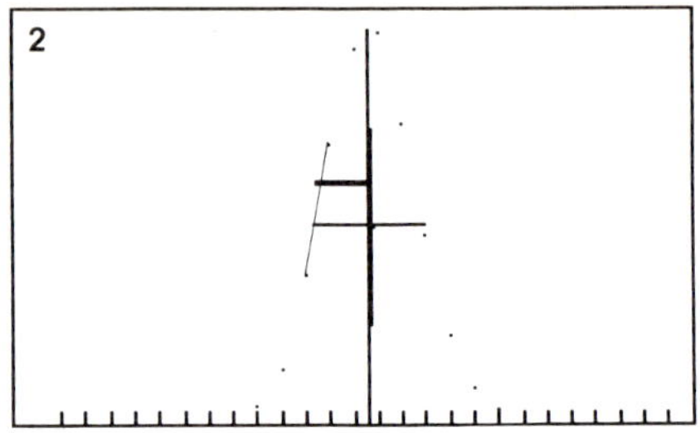

Figure 11 (peak 2)

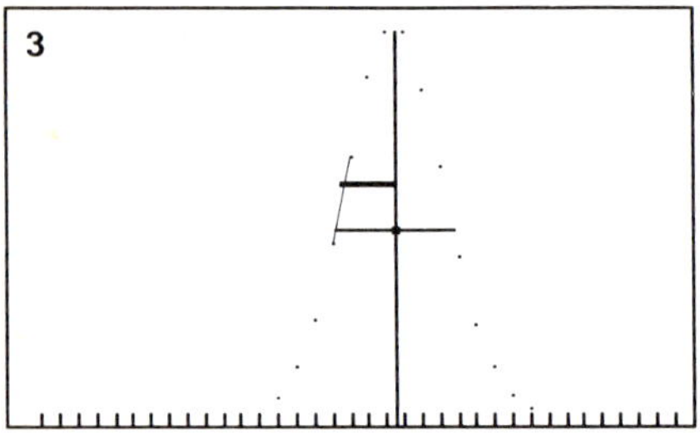

Figure 12 (peak 3)

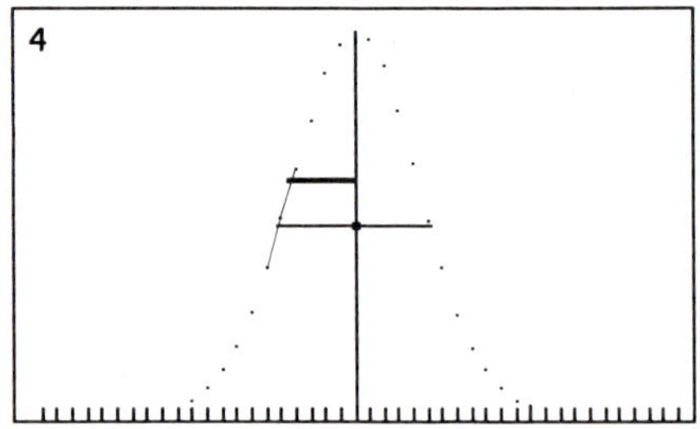

Figure 13 (peak 4)

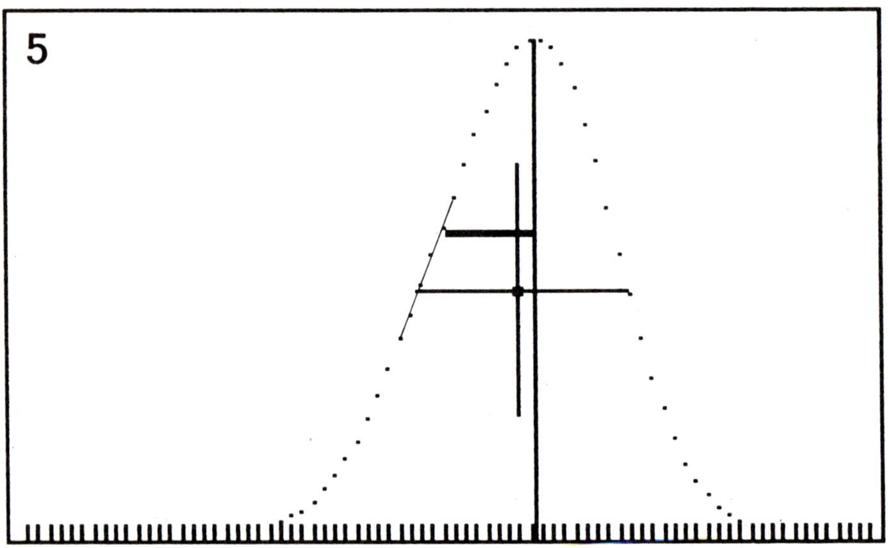

Figure 14 (peak 5)

Figure (10-14)
Raw data plot of signal over time; interpolated retention time "tms"; peak width at half height "b0.5"; "sigma" (the standard deviation as thick line); "center" as short vertical line. GC raw data set as given in Table (1).

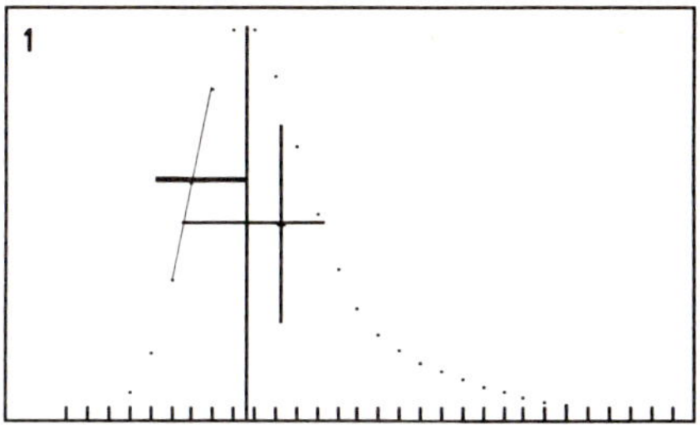

Figure 15 (peak 1)

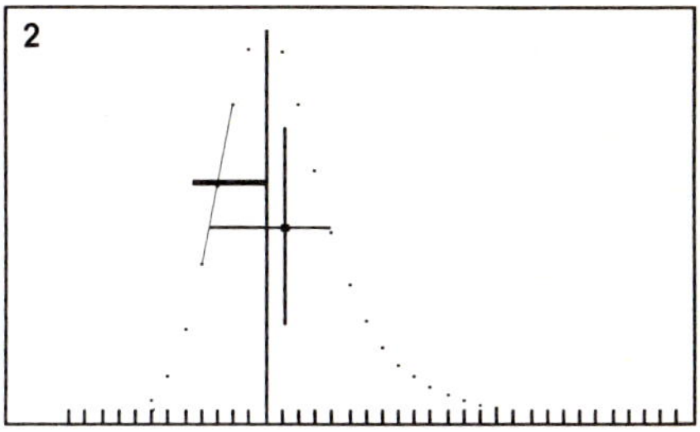

Figure 16 (peak 2)

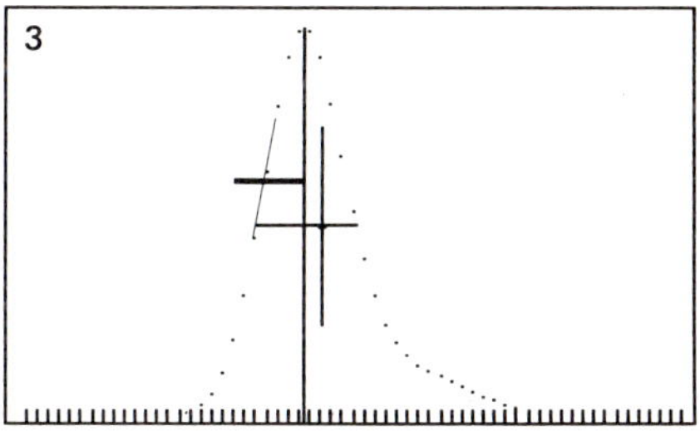

Figure 17 (peak 3)

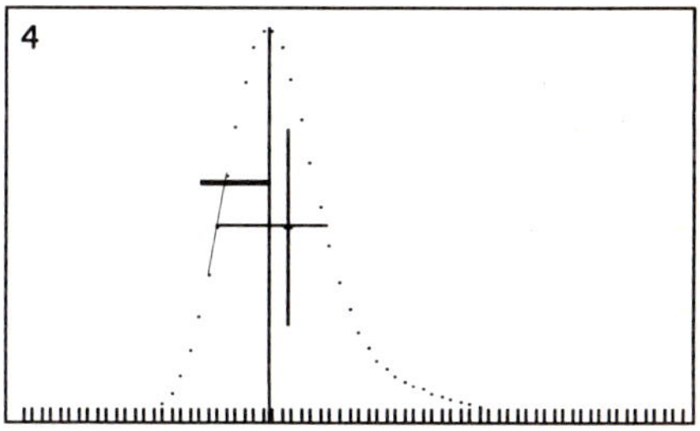

Figure 18 (peak 4)

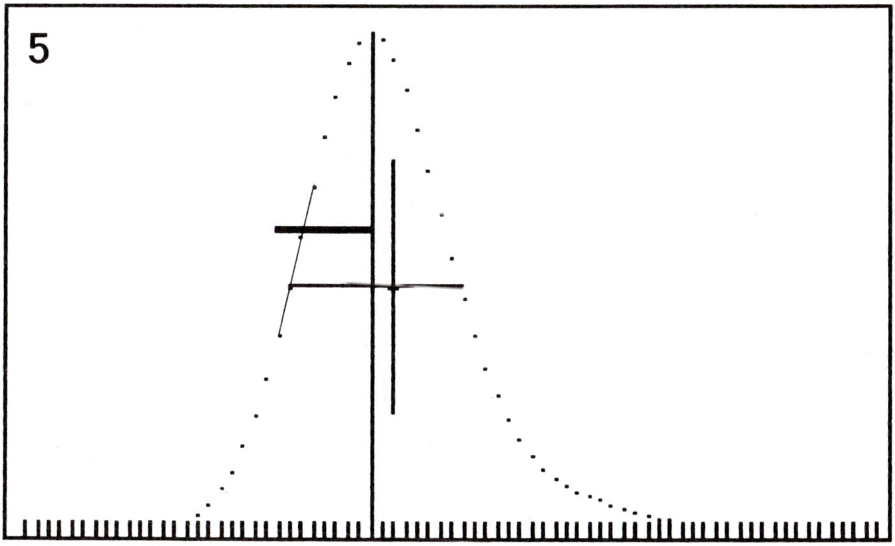

Figure 19 (peak 5)

Figure (15-19)

As in (10 - 14), but for a series of HPLC peaks. Note the remarkable deviation from Gaussian shape as seen in the far too large value for the standard deviation (s)—thick line at 60.7% of peak height.

time position of the perpendicular from the (interpolated) peak top to the baseline and the point of intersection of the line through the center of gravity with the baseline. The latter line divides the peak area into equal parts.

3) Data density increases exponentially with the homologue number of the substance chromatographed.

4) At least 5, or better 10, raw data points over the peak width at half height are needed to produce results which at least **look** profound. However, are they qualified?

To check this we use experimental mathematics plus graphics display, a most elegant tool for the chromatographer who has had only modest training in the use of mathematics in his work.

Let us interpolate the few raw data points of GC peak "one" named DA1.PK1 and increase the data points by a factor of five. Then we will again analyze

sigma
center of gravity
peak height
retention time by interpolation

and compare the data.

We will use the cubic spline as providing the best approximation to a chromatographic analog signal. Interpolation will be performed with 300 points over the entire peak range. We will calculate the cubic spline-interpolated signal data using 3 to 4 times more data points. On comparing the "standard deviation value" of peak one with only 3 to 4 raw data over the range of peak width at half height with 3 to 4 times **more** raw data we will see the improvement of the accuracy of peak shape analysis data.

10 data points across the peak width at half height will prove to suffice. This is the optimum **easily** attainable just by using cubic spline interpolation techniques if there is a lack of data points—this very provocative statement is sure to elicit protest. However, we have tested this interpolation technique with hundreds of differing shapes and found that it works, provided that the signal height data are reliable, in other words free from noise!

Moreover, this is quite feasible in practice, although not automatically guaranteed—see peak 5 of the GC peak series.

Let us first consult Figure (20) with the raw data points connected linearly and the peak replotted in very narrow form on a compressed time axis, compared with Figure (21) which shows exactly the same raw data replotted with the maximum possible peak width.
The reader will realize:

a) The graphics display of only 260 by 160 points available, say with an Apple II, is so bad that it creates an impression of lost precision and accuracy.

b) One is easily misled by manipulation of peak shape x axis values. A dense time axis gives the impression of nice and sharp peaks with no tailing. However, on stretching the time axis while leaving the y axis unchanged one starts to see peak shape deformation. Compare now Figure (21) with Figure (22); it looks "fine" again.

Let us now undertake shape analysis with the data of this "splined" peak (22) and examine the graphics (Figure (23)).

Result: The standard deviation has changed. The very large s value in DA1.PK1 (peak "one" of the GC run) was systematically falsified due to an insufficient number of raw data points. It has now been corrected, and shows similar values when compared with peak two...five of the original shape analysis. We have now 12...13 data after cubic spline treatment instead of only 3...4 prior to this legal mathematical "data massage" step.

Let us now compare shape analysis data of GC with HPLC peaks: the far larger standard deviation values "s" seen in the latter case are not the result of any lack of data but of the very distinct deviation from Gaussian shape.

Thus all discussions of **variance** in theoretical, practical, and instrumental chromatography are of little use if they are based upon the "sigma" values many chromatographers **measure** at 60.7% peak height (where b would be 2 sigma) or *via* b0.5 at 50% peak height (where sigma would be 1/2.35 of b) or at the peak base by construction of tangents. All those "sigma" data are wrong by 10 to 50% or more, and the VARIANCE values ($v = s^2$) would

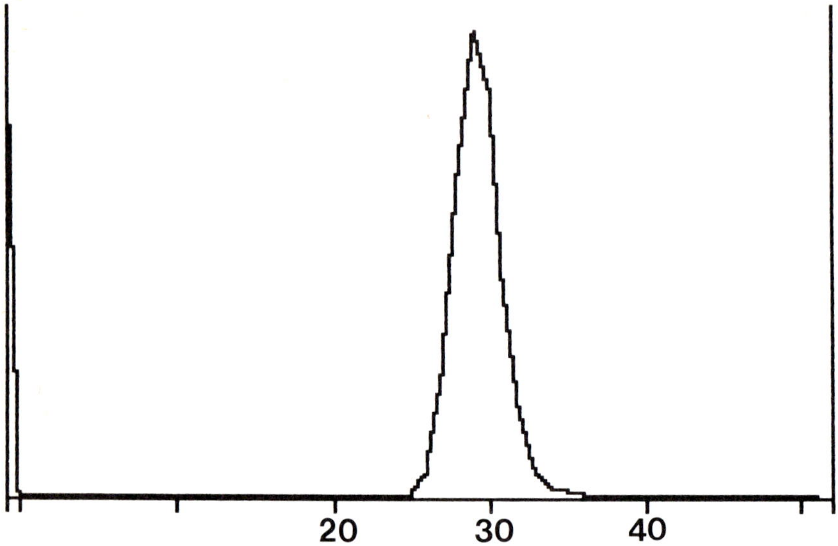

Figure (20)
"Screen replot" of a narrow (fast) capillary peak. Ticks enclose a 10 raw data range.

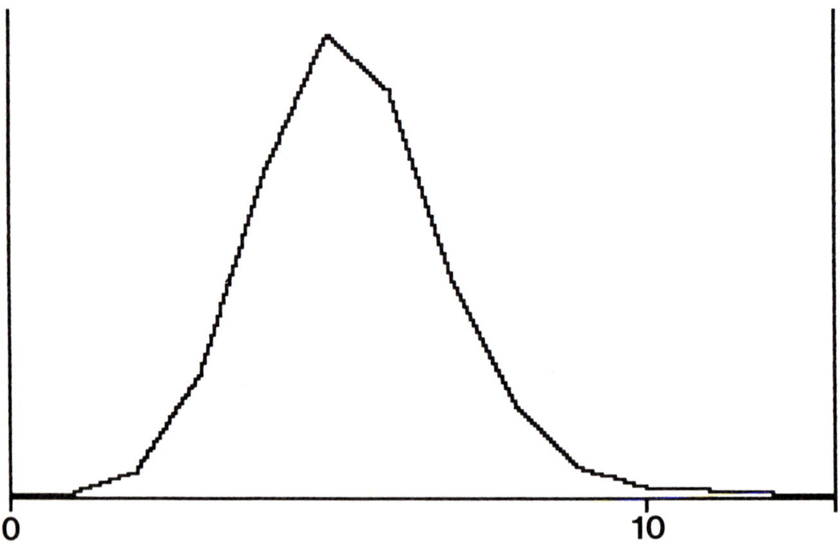

Figure (21)
As in Fig. (20) but with larger x-axis units, clearly demonstrating the insufficient density of raw data.

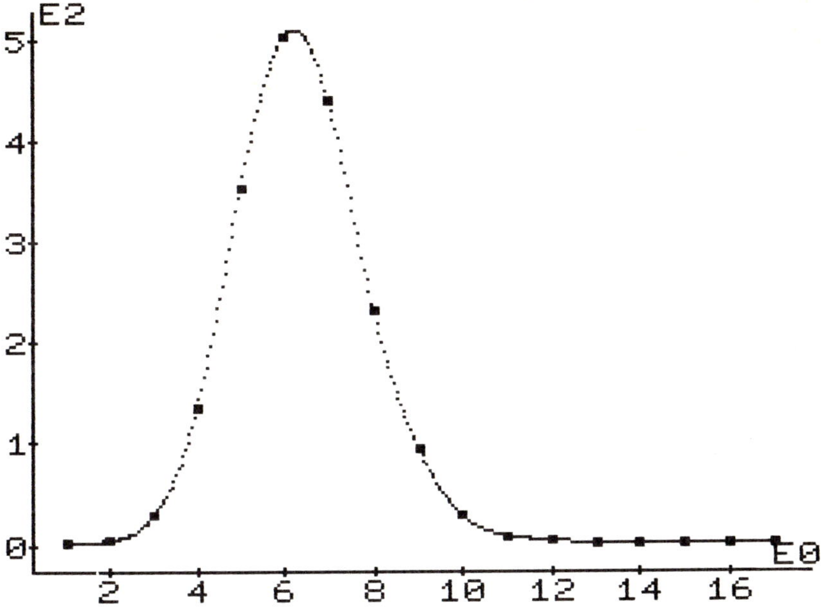

Figure (22)
Data in Fig.21 have been improved by cubic spline function—perfect peak shape, no deformation visible.

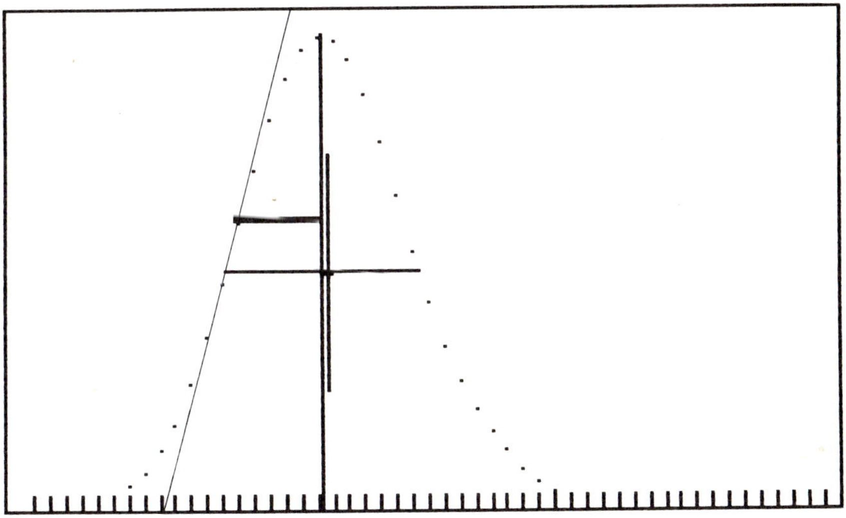

Figure (23)
Shape analysis of "cubic spline"-interpolated raw data of Fig.21 (raw data—see GC peak 1 in Tables (1...3)). Result: Value of (s) now acceptable, but not according to a true Gaussian shape.

be useless as they are systematically falsified. Anyone who thinks in terms of variance additivity calculations (very questionable under many chromatographic conditions) MUST use mathematically "clean" peak shape analysis data. The true variance is easy to calculate and the data are accurate enough if more than 10 raw data are available within the range of the peak width at half height or can be "constructed" by the cubic spline technique. However: even my wife protests against this "data massage" technique and is in agreement with many others who have doubts about the results.

Let us therefore use peak 5 of the HPLC peak series and treat the data in this way: First we will use only one third of the raw data available (see Figure (24)) and put them into the shape analysis algorithm. Then we interpolate by the cubic spline technique to increase the diminished raw data set fivefold (see Figure (25)) and use these data for shape analysis. Then let us make our comparisons

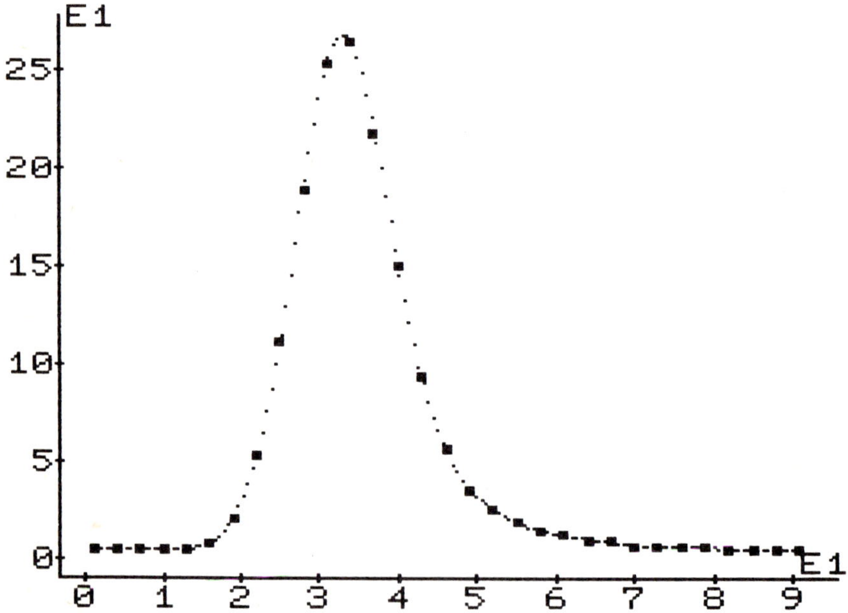

Figure (24)
Reduced raw data set (one third) of the HPLC peak no.5.

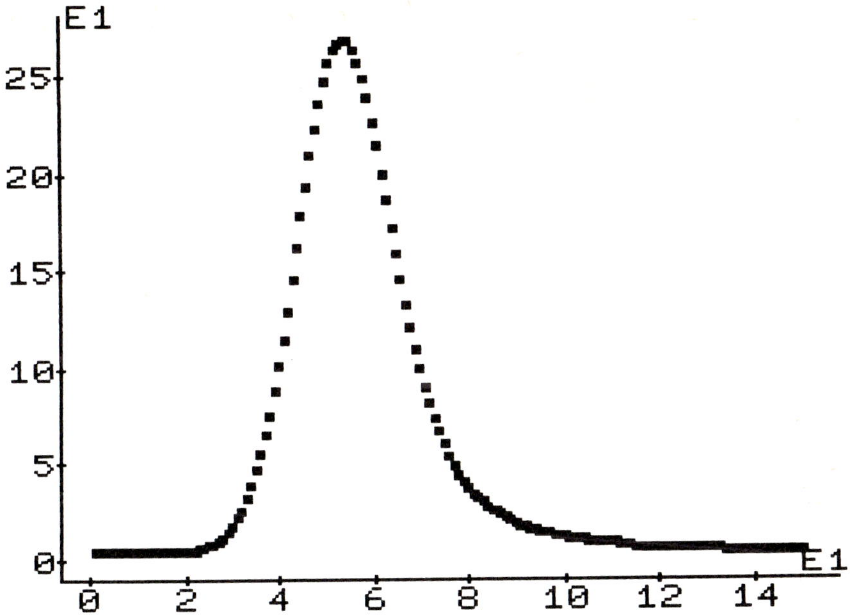

Figure (25)
Fivefold increase of the data in Fig.(24) using cubic spline interpolation.

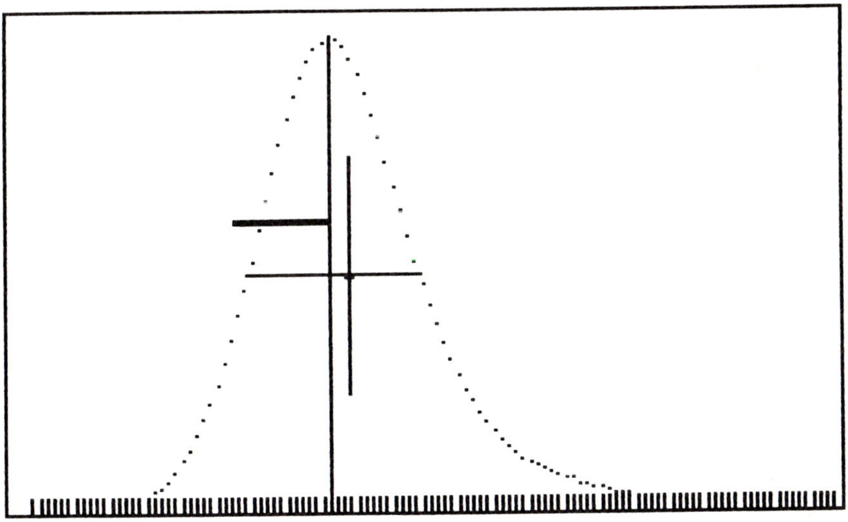

Figure (26)
Shape analysis of first reduced, then fivefold increased data set in HPLC peak 5; to be compared with the original in Fig.(19).

by means of graphics and numbers. Finally we check the results:

Relation of b0.5 to s with	5 data above b0.5	1.43
	13	1.79
	23	1.80

One could easily find how many data are essential per half peak width in order to get sufficiently accurate shape analysis data for a **minimum** of raw data to be digitized, stored, handled, and concentrated.

As a rule of thumb: 10 is adequate and supplementing data sets of sparse density by the cubic spline approach—**the** chromatographers's interpolation technique—is legal. Just compare Figure (19) with Figures (24) and (26).

Sigma (or s data) or variance data are **not** so important. Much more valuable are the real peak width values at 15%, 50%, and 85% of peak height. As chromatography is a statistically "bad process" subject to disturbance (with regard to peak shape) by the injection and detection steps (*Golay* [9]), it is appropriate to recall the proposal of *Landau* [10]: the best measure of signals from statistically bad processes is the peak width at half height. To this may be added the peak width at 15% and at 10% of peak height; these two values should be divided into the part ahead the peak maximum and that behind it. The relation of the component of b0.15 AND of b0.10 at **higher** retention time to the corresponding components at **lower** retention time would provide an easy and sufficiently accurate quantitative measure of **tailing.** Cubic spline interpolation of the peak shape curve at 15% **and** 10% height, left and right of the maximum, would furnish accurate data.

Here is an example for the b0.15 value taken from one of the single peak raw data in the raw data table. ("factor" in the following means: time steps, where raw data have been taken, or the sampling rate expressed in time units, see the ticks in one of the figures around Figures (17...19):

At tms = 32* factor is the peak maximum of H = 264.6.

At t* factor = 20.37 and 46.57 we find the peak to have reached 15% of its maximum height. So the "tailing factor" TAI = (46.72 − 32)/(32 − 20.37) = 1.27 at 15% of maximum peak height. The procedure is repeated with the data at 10% height. If the tailing comes from physical sources, then it should obey the exponential dilution law

$$H = H(max) * \exp(-V/F) * t$$

where t is the elution time
H(max) is the peak height
F is the carrier gas flow
V the remixing dead volume $\ll$ system dead volume
and H is 15% and 10%, respectively, of the signal level.

Does the TAI factor from the 15% height level increase to TAI at 10% height level according to the exponential dilution law and is this true for ALL peaks? If the answer is yes, the reader should know where to reduce (by mechanical improvements) the over-large tailing factor. This is just one little illustration of what can be done with peak width data!

Let us now use the most important value among the peak width/peak shape data: the peak width at half height b0.5.

It is no longer appropriate to work with "sigma"; instead, it is correct to work with "s values". However, having analyzed thousands of peak shape families in GC and HPLC under widely differing conditions for many years, we know that there is no simple underlying mathematics. There is no hope of finding a class of **shape** laws which would accommodate most of the shapes encountered in chromatography. In this situation, we should concentrate on what we know to work: this is the simple b0.5 value—which **cannot be measured precisely enough.** An illustration will immediately illustrate the meaning of this repeated statement:

Values of plate height, plate number, separation number, *etc.* depend on b0.5.

How accurate will for instance a "plate number" be if the peak width at half height is about one second long and the time data is collected with a precision of ±0.1 second? Just an example:

b0.5 = 1.1; tms = 30; N(theor) = 4100
b0.5 = 0.9; tms = 30; N(theor) = 6200

5.3 Peak Width at Half Height Correlated with Time

or k = (tms − tm)/tm

We will end up with data describing in full all details of the chromatographic system except the mobile phase/substance/stationary phase interaction. To start with we need precise data about the system being analyzed.

The GC data have been stored in a "CONDITION" file.

Naming convention: the system data corresponding to any system analysis are named "C.(name)" and all corresponding information—see list below—are stored (as normal readable text) in a C.(name).COND file.

This includes:

Sample type; volume; temp.of inj.port; sampling mode; stationary phase; sep.mode; temp.mode; mobile phase; flow; flow mode; inlet pressure; outlet pressure; spec.info; starting temp; end temp; heating rate; detector type; det.sens.; det.mode; det.conditions; length of sep.syst.; particle/inner diameter; if particle: distribution analysis; if open tube: i.d.; AD converter data AD rate; AD range; AD mode; add.remarks; who was analyst; instr. specialities; *etc.*

It is simple to run a machine language search to combine C.(name).COND files together with the raw data file and to organize raw data analysis programs so that they automatically check for an available C.(name). COND file; read the content; put it into memory; use it for system analysis and preparing a report; and give an ALARM signal if anything lies far outside the expected or normal range.

Upon examining the C.(name).COND file we find:

stat.phase SE30
mobile phase nitrogen
inlet pressure 0.1 bar
sep.system fused silica capillary
length 14000 mm
i.d. 0.28 mm
isothermal 100 deg. C
detector FID
sens 1 x 1 (full)
sample test mixture hydrocarbons C6...C10 in C11
volume 0.2 microliter
and most unexpected (hopefully no wrong input)
AD sampling rate 0.77 s !

We start the program SYSTEM analysis.

It reads the raw data cut into peaks, concentrated to optimum retention time raw data, peak width at half height, and s data (the approximation value of the standard deviation), as well as the retention index or molecular weight data for all homologue peaks. Nothing more is necessary for the enormous number of results. All these data are serially concentrated in a "refined data file" with the naming convention

C.(name).PK

Now as the D/A conversion rate of 0.77 s could be used to recalculate all data in seconds, we obtained the following Table (7) of data:

Table (7)

peak no	tms (s)	b0.5 (s)	s (s)	variance error
1	266.73	2.68	1.84	20.0 %
2	353.89	3.54	1.66	9.3 %
3	520.91	5.08	2.33	8.9 %
4	839.38	7.92	3.54	8.5 %
5	1447.99	14.72	6.29	6.0 %

Variance error is the deviation of the value found for s/peak width at half height from the theoretical value in case of symmetrical Gaussian-shaped peaks, which is sqr(8∗ln(2)) = 2.355.

The next value calculated by a "chromatography system analysis program" is the dead time tm, now given in seconds. It was found to be

tm = 170.73 s (B = 0.99999987)

5.4 Theoretical and Effective Plate Number

On the basis of this information the program calculates the k value of each peak.

This made the next Table (8) of data possible:

Table (8)

no	ts (s)	k	index	N(th)	n(eff)
1	96.49	.56	600	1108	144
2	183.64	1.07	700	1955	525
3	350.67	2.05	800	4233	1915
4	669.14	3.91	900	10984	6972
5	1277.75	7.48	1000	32673	25425

The reader may be shocked by these very bad "theoretical" and "effective" plate numbers? Well, in fact they may be considered as bad decision numbers anyway, because they are highly dependent on how the chromatographic system is **mechanically** coupled together. If the mechanical "error" is large **and** the mobile phase flow large then b0—the starting peak width (see below)—is large. And a large value of b0 will amplify the influence of the "chromatographic position" of a test compound. Meanwhile, more and more chromatographers are realizing that the "holy cow" of theoretical or effective plate number and the related plate height values are to be used

a) either at a given k value or

b) on additionally stating the k data with the plate number/plate height data.

Otherwise these numbers are devoid of any real value.

Now most of the publications on column "preparation" (yet too few on instrument construction and optimal column use) assign a prominent role to the "HETP curve", which shows how the plate height changes with the mobile phase flow rate. Yet even the HETP/u minimum is of no real practical help in optimizing a chromatographic system, as in fact there is MORE than one optimum value for the phase flow. This disaster is one of the reasons for the small number of optimized systems encountered in practice.

In order to avoid such discussions, some members of the chromatographic community have decided to use all the plate numbers at k = 0.4—see Ref. [11]! See Figure (29) to understand, that our colleagues' decision is most questionable! A few other colleagues simply stated that the concept of precalculating all height—and number values for all k or tms data are wrong and valueless—one of them see in Ref. [11]. On the other hand, however, more and more colleagues are using computers in the lab. They calculate their numbers, see the "side effects", and agree more and more with the so called "ABT concept", which is our name for the procedure we developed and have been using in practice (and by far not only we) to great benefit for a decade:

5.5 "b0.5 = k*a + b0": Slope and Intercept; Practical Use of "ABT Concept"

We examine the correlation of peak width with time. If there is a simple correlation (as there is in most practical cases) then we find the quantitative dependence expressed in an easy to understand formula:

b0.5 of any peak P in isothermal GC = k * a + b0

k = (tms of this peak P − tm)/tm
a = slope
b0 = intercept of the peak width *vs* k regression line

in the case of linear correlation.
If the correlation is non-linear, then we must use an expression for the dependence of b0.5 upon k of the following kind:

$$b0.5 = a0 + a1*k + a2*k^2 + a3*k^3 + a4*k^4 + a5*k^5$$

but **cannot** extrapolate any peak width value.

In order to reserve the possibility of extrapolating data, we prefer to use **linear** correlation and are prepared to live with the limitations.

This Section will show what the limitations are and how easy it is to misunderstand the "ABT concept", or to reject it too early. We call the correlation of peak width data with time, when time is expressed as k = (tms − tm)/tm the "ABT concept". It is of great practical help, once understood. In fact it is the basis for "auto-optimization" of a separation process in GC and HPLC. This will be shown in the following volumes of "Computer Chromatography". Here we will illustrate the practical usefulness of the correlation "peak width/time" by some data. The reader is therefore invited to have a look to the NEXT data set, which follows the one we have used exclusively up to now for GC. So please be patient, some colleagues misintepreted the ABT concept too early, realizing much too late with the benefit of considerable hindsight that it says no more than how peak width changes with k, flow, temperature. But that IS precisely what separation is all about. It is no more and no less than a description of the dependence of the peak width at half height on the rest of the chromatographic world. We are just correlating peak width with time.

There follow the next valuable data:

● Test for the correlation of peak width and k. If linear enough one could interpolate any plate number, plate height, or separation number. See Figure (27).

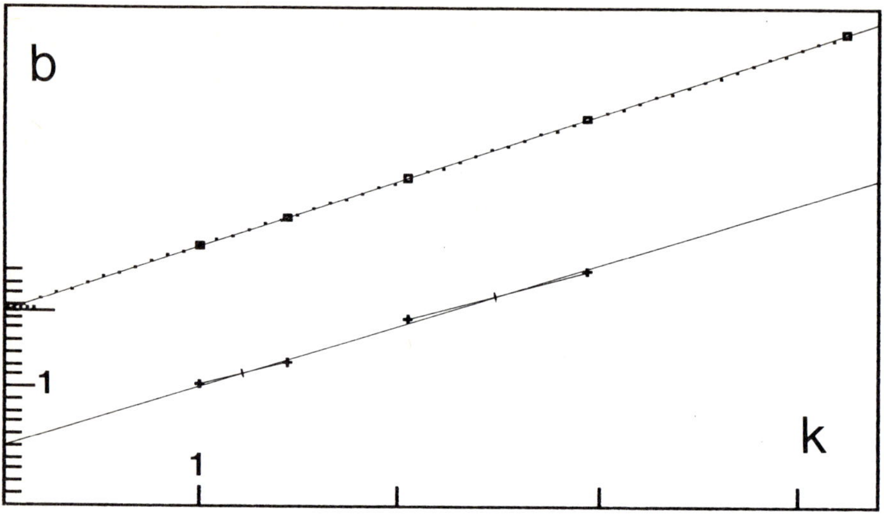

Figure (27)
Astonishing linearity between peak width at half height "b0.5" = Y axis and k = (tms − tm)/tm on X axis from an optimized capillary GC data set (not listed in this volume; see Computer Chromatography, Volume 2, Optimization).

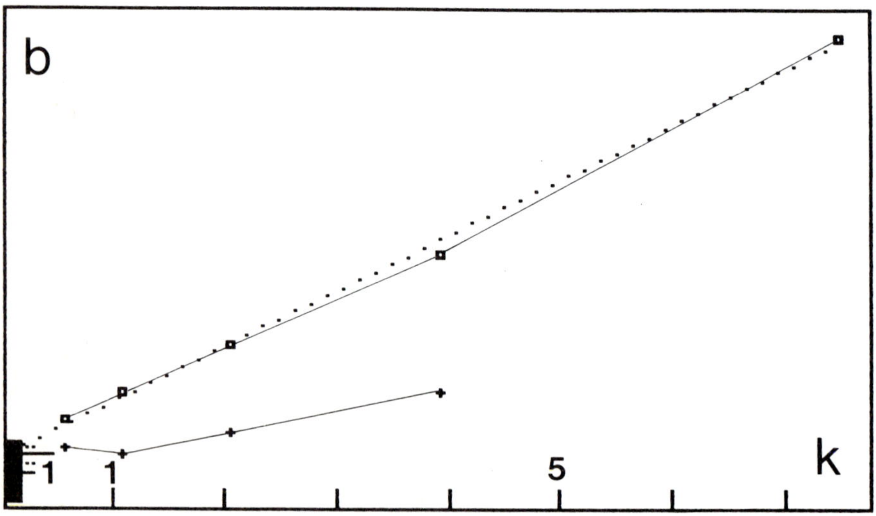

Figure (28)
Not too good a linear correlation of "b0.5" with "k" on X axis and no linear correlation of "Sigma" = + + + with "k".

- Inter- and extrapolate analysis time and molecular weight. This will furnish a complete set of information about the working range of a separation system under the conditions used (pressure, flow, temperature).

- Compare system-relevant data with data known to correspond to technical or theoretical limits. This will tell you "how far you have come" with your tool.

We have concentrated the data intercorrelation into three basic values:

into a = peak width increase in seconds per k unit
b0 = starting peak width in seconds,
extrapolated for k = zero, *i.e.* where chromatography starts. If b0 is not zero seconds, but some 5 seconds, any peak will be 5 seconds broader than necessary from the standpoint of the "physics" of chromatography.

Figure (28) shows that a second order interpolation might fit the data better? Well, we want to **extrapolate** a large number of values in order to acquire sufficient knowledge about the range of data within which we work with our chromatographic system. And extrapolation becomes useless very soon.

Nevertheless, let us try it. A laboratory computer is a good tool for mathematical, or at least for "computational", experiments.

A second degree polynomial regression gives the following data as compared with first and third degree regression:

degree	regr. qual.	determ.coeff
1	3.2 ± 2.1 %	B = 0.997
	(b0.5 = 1.57	+ 1.73∗k)
2	1.53 ± 1.13%	B = 0.9997
	(b0.5 = 1.98	+ 1.36∗k + 0.045∗k²)
3	0.04 ± 0.04 %	B = 0.999999
	(b0.5 = 1.69	+ 1.83∗k − 0.12∗k² + 0.013∗k³)

Third degree regression through 5 points too often looks too good. In data analysis at least, this good correlation is too often just the result of systematic **errors** and is rarely repeatable.

Theoretical models tell us that in GC peak width changes with time according to second order functions which very closely approach first order correlation. In particular, substances of low molecular weight eluting early at low temperatures—for instance methane—often have broader peak widths than the next eluting substances. But this **is of no importance** if we are interested in correcting peak width data by subtracting that part of **additional** peak broadening which is independent of chromatography and diffusion constants, but which is induced by extra-column effects from the sample inlet through injectors, connecting parts, down to the time constant of the signal cable resistance, recorder, or integrator.

Would you believe that an absolutely minimal **extra-column** induced peak width of a few milliseconds can result under one set of system conditions while a 6000 millisecond **additional** peak broadening obtained under another, bad set of chromatography conditions (but with the very same column)! This is the reason why we undertake system analysis—to avoid killing the column instead of eliminating the real source of trouble.

In HPLC most isocratic runs of test substance mixtures within homologous series produce peak width *versus* k correlation of B = 0.9999 — linearity, too linear for us to use a second order correlation, see below.

Back to our GC test data:
Whether 1st, 2nd, or 3rd order regression, there is a **b0 value** between 1.6 ... 1.7 ... 1.9 s width in this set of data. But b0 is extremely sensitive to the smallest systematic errors of measurement and very strongly dependent on mobile phase flow at low speed, and although second order correlation would often be the better model we stick to the linear formula

$$b0.5 = a * k + b0$$

for the reasons mentioned.

The formula given above, however, is incomplete. The peak width **slope factor "a"** as well as the intercept **b0** have standard deviations. Too blatant a misuse of linear correlation would be detected on examination of the additional data "sa" and "sb0", *viz.* the two standard deviations.

The program did in fact calculate the key data:

a (in seconds peak width increase with k) ± rel. standard deviation and b0 (in seconds, extrapolated starting peak width at k = zero) ± rel. standard deviation

and found

a = 1.72 s ± 2.9%

b0 = 1.56 s ± 12.9%

Anybody who knows about the practical use of correlation data and the danger of any extrapolation if not truly linear will have misgivings regarding the use of such data (with b0 of ± 12% uncertainty).

Although we have used AD conversion of 12 bit resolution, although we worked with raw data storage and error-free (foolproof) programs, chromatography system analysis really is **not** easy, especially in the case of capillary gas chromatography.

So to repeat and to repeat again is the "*conditio sine qua non*". (Stated simply: Perform more than one system analysis. Just one is no use at all.) Analysis should be repeated after several days using a test sample with larger homologues to avoid critical small values:

We kept instrument, column, mobile phase, pressure, temperature constant. We changed the sample from C6...C10 to C8...C12 hydrocarbons. This gave larger adjusted retention time data. The DA-conversion rate was 0.755 seconds. The first to last peak all had sufficient raw data per peak (all above 10 measurements over the peak width at half height).

Result of system analysis:

A	B	T
a = 1.61 s ± 0.5%	b0 = 1.91 s ± 1.1%	tm = 175.7 s

Linearity of peak width *versus* k : determination coefficient B = 0.9999. See the graphics:

Figure (29) shows perfect agreement between simply calculated discrete plate numbers and the interpolated line of the dependence of the effective plate number over k = (tms − tm)/tm and the theoretical plate number N over k.

Figure (30) shows how well b0.5 data of other homologues may be extrapolated.

The fact that the dead time calculation resulted in a determination coefficient of B = 0.99999999 supports

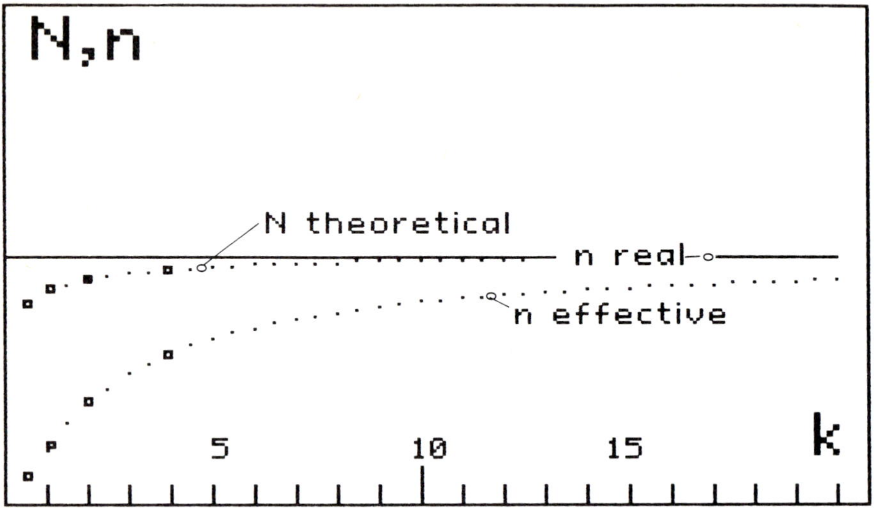

Figure (29)
Theoretical function of N = theoretical plate number and n = effective plate number over “k”
Value of n(real) = line in middle of the Figure. Correlation between theoretical and measured data: see position of the small open squares.

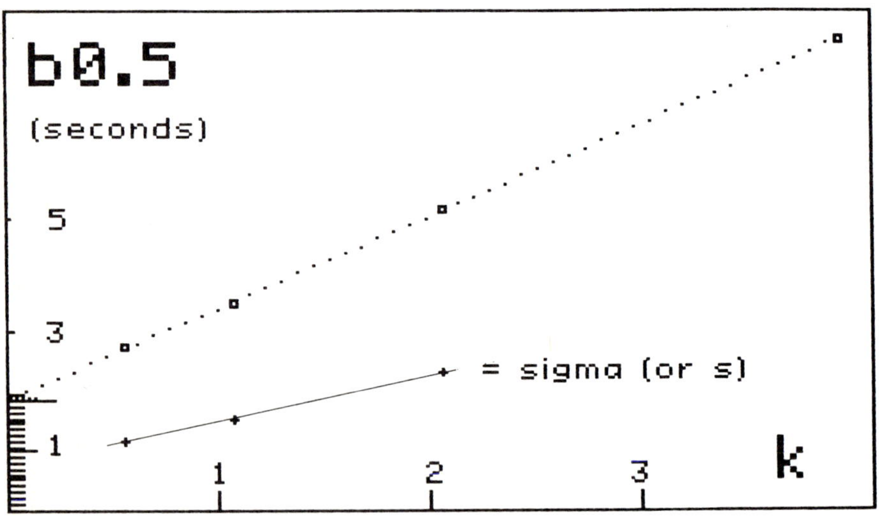

Figure (30)
Peak width and “k” correlation in isocratic HPLC. Note the large extrapolated “b0.5” value at “k” = 0!

the idea of calling for inter- and extrapolated numbers for other homologues, or substances of very similar chromatographic behavior.

Might it not be of interest to see in which range the column just analyzed has **no** resolution, or how long the retention time of, say, C20 will probably be? If tms C20 is shorter than an hour, why not use this column or capillary which has just been system-analyzed for samples in the range up to C20? However, if C20 is eluted only after 10 hours retention time, nobody will then be prepared to sit at the chromatograph waiting for the end of the analysis! Most chromatographers wait a lot of time for the end of many runs and would **not** wait—or better: would **not even inject** a sample—if they knew beforehand that elution would take a week.

5.6 BASIC Algorithm for Forecasting k, N(theor), n(eff), TZ

These are the algorithms written in BASIC for forecasting in gas chromatography:

1) Find the slope AN and the intercept BN of the linear regression "log(tms − tm)" over the carbon number

2) Find the slope A and the intercept bo of the linear regression peak width (sec) over k = (tms − tm)/tm.

Start interpolation with the carbon number C1 and end with the carbon number C2

```
100 FOR I = C1 TO C2
110 K2 = AH * EXP(AN*(I+1)*100+BN):
    K1 = AR * EXP(AN*I*100+BN):
    K2=K2/TM:
    K1=K1/TM
120 K(I)=AR*EXP(AN*I*100+BN):
    K(I)=K(I)/TM
121 REM K(I) is the k value of each homologue
130 NT(I) = 8*LN(2)*((K(I)+1)*TM/(A*K(I)+b0)—1
131 REM NT(I) is the theoretical plate number for each
    homologue
140 NE(I) = NT(I)*((K(I)/(K(I)+1))^2
141 REM NE(I) is the effective plate number for each
    homologue
150 TZ(I) = (TM*(K2—K1))/(A*(K2+K1)+2*b0)—1
151 REM TZ(I) is the "TRENNZAHL" between the
    homologues
```

```
160 PZ(I) = (LOG((K(I)*A+b0)/b0))/LOG((TM+A)/
    (TM-A))-1
161 REM PZ(I) is the integrated number of baseline
    separated peaks within the range of
    homologues chosen.
170 NEXT
```

If you want to have the data in ANY range of the chromatogram, at discrete "time data" then use the following algorithm or just use a FOR...NEXT loop starting with k = 0 and stepping in any of the step width of k = (tms − tm)/tm you want OR for a single value "k": Calculate "k = (tms − tm)/tm" for the position in the chromatogram chosen.

NT (N theoretical) = $8*\mathrm{LN}(2)*((\mathrm{TM}*(k+1))/(A*k+b0))^2$

$k = (tms - tm)/tm$

NE (n effective) = $8*\mathrm{LN}(2)*(\mathrm{TM}*k/(A*k+b0))^2$

You cannot define a "Trennzahl" for any chromatogram position, but there are other valuable data describing the separation efficiency at any given position in a chromatogram.

There is one plate number, which is independent of k(substance) and b0:
the real plate number n(real):

n(real) = $8*\mathrm{LN}(2)*(\mathrm{TM}/A)^2$

You will see if n(real) is equal to N(theoretical): this is when the slope of the increase of peak width over k equals the b0 value. This is possible under certain experimental circumstances. In those cases N(theoretical) remains constant throughout a whole isocratic chromatogram.

On inspection of this formula it is immediately understood that the "Trennzahl" depends on the range of the chromatogram in which k1 and k2 are eluted; however, there is a "Trennzahl max" in isothermal GC above which TZ is a constant (which is in total disagreement with the statements of the authors in Ref. [13]).

Up to here all separation or column-characterizing values are dependent on k **and** dependent on b0. The former is substance dependent, and the latter depends on how the chromatographic system is used, installed, which technique of installation, *etc.*

We therefore searched for quality numbers describing separation efficiency independently of the specific substances chosen. It is possible to use "Trennzahl" for this purpose if one calculates the final maximal possible value which is independent of the substances examined. However, TZ(max) will depend on b0, as all chromatographic efficiency depends on this value which is of essentially mechanical origin. "Trennzahl C19—C20" is easy to understand: a TZ (19—20) = 60 means that a maximum of 60 (uniformly distributed and consecutive) peaks can be baseline separated between the two homologues C19 and C20. And as this covers a range of a hundred retention index units, one also knows which substances characterized by retention index can be just baseline separated. This is d I = 100/(TZ + 1) which is in our example

d I = 100/61 = 1.64 index units difference.

If you now have some feeling for how long the column must be to achieve a Trennzahl of 60, you will find that this is at least 30, but more probably 90 meters, resulting in the absolutely indispensable nature of "capillary GC".

After many years of testing in our daily work, we finally introduced the separation number SN 0—10. Here we calculate the number of baseline separated peaks using a "hypothetical" substance with k = 0 and another borderline substance of k = 10.

In BASIC the algorithm is

```
SN = log((10*a+b0)/b0)/log((tm+a)/(tm-a))-1
```

If you want to express the numbers of peaks one can separate between k = 5 to k = 25 then this is

```
SN (5-25) = log((25*a+b0)/(5*a+b0))/log((tm+a)/
(tm-a))-1
```

and if you want to know the separation capacity in peaks which are baseline separable between the start of the chromatogram and k = 30 then

```
SN (0-30) = log((30*a+b0)/b0)/log((tm+a)/(tm-a))-1
```

This separation number is highly flexible, but k is not all that good for forecasting.

We have therefore used the excellent linear correlation between the retention index and the log(adjusted retention time) in isothermal GC. There follows the next linearity formula of practical value:

I = a∗(log(k∗tm)) + b0 and

exp((I−b0)/a) = k∗tm

One can now replace k by this exponential function and can forecast everything in any range of carbon numbers (carbon number of *n*-alkanes equals I/100), provided, of course, that one is convinced of the effectiveness and validity of linear correlation!

Up to now, in strictly isocratic GC, we have never found any critical deviation between forecast and experiment within a few days.

In order to provide the reader with some quantitative inter- and extrapolated data we have used a "peak shape auto analysis program" with the raw data of a gas chromatogram, where C8...C11 hydrocarbons in C12 as solvent have been separated at 100 °C, on a 14000 mm long capillary with SE30 silicone rubber as stationary phase.

The shape analysis program cut the raw chromatogram into the peaks, interpolated for the optimum retention time, and calculated the peak width at half height. The data set was completed by retention index data (800....1100)

Results as REPORT PRINT OUT: see Table (9).

The "maximum TZ" for this separation system is TZ(max) = 32.8, which in fact is not bad for a 14 m long capillary of 0.28 mm width.

The capillary has at tm = 175.7 s quite a good
real plate height h(real) = 0.21 mm
n (real) = 65500 (for 14 meter !)
SN 0−10 = SN (real) = 121 peaks

but the analysis speed is low:
SNt, which is the mean number of peaks eluted within one minute was

SNt = 3.8

The summary of all these quantitative data is concentrated in and given by the three basic values, the "ABT" data

a = 1.62 ± 0.009 s
b0 = 1.91 ± 0.02 s
tm = 175.7 (B = 0.9999997)

```
 DATA SUMMARY TO C.DA1.1.PK

***  USED PEAK DATA

TMS        B0.5     SIGMA    INDEX   K'    VARIANCE ERR.(%)

275.801    2.854    1.185    800     .56     -4.5
366.175    3.639    1.578    900     1.08    4.1
538.994    5.262    2.378    1000    2.06    11.7
867.872    8.282    3.647    1100    3.93    7

***  SYSTEM DATA

 LENGTH OF COL./CAP.:             14000 MM
 I.D./ PARTICLE-D. :              .28 MM
 STATIONARY PHASE  :              SE30
 TEMPERATURE       :              100 DEG. C
 PRESSURE          :              .2 BAR WITH N2

***  ABT DATA

*A(SEC) = 1.62       *B0(SEC) = 1.91       *TM(SEC) = 175.713
 DETERMIN.COEFF. OF DEAD TIME ITERATION          : .999999743
 DETERMINATION COEFF. OF TMS/B0.5 CORRELATION: .999
 STANDARD DEV. OF A (SEC)     = 9E-03 ;    OF B0    = .021

***  SEPARATION NUMBERS

 SN REAL = 121          N REAL = 65529.52   H REAL = .213 (MM)
 SNT     = 3.75       (PEAKS/MINUTE IN RANGE K= 0 TO 10)

***  QUALITY OF PACKING OR CAPILLARY IMPREGNATION : ETA = 131 %

 K'-RANGE USED FOR ANALYSIS    : FROM K= 0 TO 3.9

             extrapolations (take care!!)

          START WITH C NUMBER    >>> C  = 7
                         BIS     >>> C  = 17

          INDEX K          TMS      TZ   N(TH)   PZ
          ------------------------------------------
          900   1       6.1  MIN   18.3 55324   34
          1000  2       8.9  MIN   23.2 58316   53
          1100  3.9     14.4 MIN   27   60901   78
          1200  7.5     24.9 MIN   29.5 62779   107
          1300  14.2    44.7 MIN   31   63978   138
          1400  27.2    1.3  HRS   31.8 64681   171
          1500  51.9    2.5  HRS   32.3 65073   205
          1600  98.8    4.8  HRS   32.5 65286   240
```

Table (9)
Report of a GC system analysis done with a test mixture C8...C11 in solvent C12 n-alkanes. The Table includes extra- and interpolations of "k"; "tms"; TZ ("Trennzahl"); N(theoretical), and an integral number of the total peak separation number "PZ".

plus the slope A and the intercept B0 of the log(ts) over Index regression.

Here are the LIMITATIONS:

All data change if tm changes.
They will also change in case of pressure gradient change. (A change in the weather is enough!)

Next they will change dramatically in the case of a temperature alteration.

This means that a full system analysis MUST be done within a meaningful range of pressure-, flow-, and temperature changes. Let us assume that we—or a clever automatic on-line computer—have finally found the optimum data for maximum separation efficiency within a useful range of temperature. In concentrated form the results would then contain the following information:

Y-axis is retention index (or molecular weight, carbon number, k).

X-axis is time (seconds) from 1 second to a full working day = 28800 seconds (1983), and on the X/Y field we will find straight lines representing the temperatures.

The log(ts) data are linearly correlated with 1/abs. temperature.

And if one realizes that even a small change of column temperature will alter the dead time quite dramatically at constant inlet pressure, one will use only correct dead time data. A good determination coefficient of B = 0.99999 and better for the regression line of log(ts) *versus* 1/T (Kelvin) will be found when using this temperature-dependent dead time. Accurate data, however, will be found only after the column temperature is at constant equilibrium value. This might need half an hour's equilibration time even with most modern equipment. (All these pieces of "hardware" and procedure information are "by-products" of system analysis with graphics display and control statistics! Quite effective in greatly improving accuracy and precision in daily routine work because one "starts to stop" believing in various manuals as far as accuracy and time constants of some instrumental data are concerned!)

Perhaps the reader may have realized by now on comparing the two GC data sets—the first badly done, the second well done—that something must be wrong with the carbon numbers given in the test samples? If this is the case, he or she is quite right: it **is** useful to use k values for a critical check of which compound is where in the chromatogram. log(k) also correlates linearly with 1/T (T, Kelvin, absolute temperature).

This "full range analysis" is very time consuming when done "by hand" and results in a huge amount of data, which cannot be measured in the lab without on-line computer use. There is simply no time available and therefore knowledge coming only from such overall analysis is lacking. But it is precisely the routine lab which needs optimized conditions, because it is here that time, repeatability, and accuracy are at a premium, and the last mentioned aspect may actually be even more cost effective in practice.

Is this high information yield from a few well controlled data obtained for a homologous series also available in HPLC? Yes, it is.

5.7 System Analysis Data with HPLC Homologous Series

One data example: We used exactly the same software for the full analysis. At the Institute for Chromatography we consider some slight drawbacks of our ABT concept as of no importance when compared with the facts of common language, definitions, measurements, and algorithms in both chromatographic elution techniques: GC and HPLC.

The final data are quite different, and the reader is invited to inspect Table (10).

Now we compare N(theor), n(eff) with n(real) over k and see some inaccuracy with N(theor) of the second eluted peak (homologue number 400 or C4 ester or butyl ester of 2,4-dinitrobenzoic acid).

That there might be an error could be seen in the determination coefficient B = 0.998. It should be better! When reporting about a **flow** optimization we will see true quality data measured under the hectic circumstances of a 25-person chromatographers' training course.

The results—although densely packed—will be readable by anyone who has followed the text.

```
DATA SUMMARY TO C.BI-1310/A.PK

***  USED PEAK DATA

TMS        B0.5     SIGMA    INDEX   K'    VARIANCE ERR.(%)

131.23     3.724    2.349    300     .93     54.7
158.4      4.142    2.447    400     1.32    48.4
196.02     5.088    3.371    500     1.88    58.9
250.525    6.358    3.988    600     2.68    54.2
328.405    8.102    4.537    700     3.83    42.5


***  SYSTEM DATA

 LENGTH OF COL./CAP.:          100 MM
 I.D./ PARTICLE-D.  :          3E-03 MM
 STATIONARY PHASE   :          ODS 2
 TEMPERATURE        :          20 DEG. C
 PRESSURE           :          120 BAR WITH ACN/W

***  ABT DATA

*A(SEC) = 1.54       *B0(SEC) = 2.2       *TM(SEC) = 67.99
 DETERMIN.COEFF. OF DEAD TIME ITERATION        : .999985761
 DETERMINATION COEFF. OF TMS/B0.5 CORRELATION: .998
 STANDARD DEV. OF A (SEC)    = .036 ;    OF B0    = .084

***  SEPARATION NUMBERS

 SN REAL = 44.8         N REAL = 10816.35    H REAL = 9E-03 (MM)
 SNT     = 3.58        (PEAKS/MINUTE IN RANGE K= 0 TO 10)

***  QUALITY OF PACKING OR CAPILLARY IMPREGNATION : ETA = 32 %

 K'-RANGE USED FOR ANALYSIS    : FROM K= 0 TO 3.7
```

Table (10)
Report of an HPLC system analysis done with a homologous series test mixture (2,5-dinitrobenzoic acid esters) on a reversed phase column. Content and details as given in Table (9).

```
    extrapolations (take care!!)

 START WITH C NUMBER    >>> C  = 1
                 BIS    >>> C  = 10

INDEX K'         TMS        TZ    N(TH)   PZ
-----------------------------------------------

200    .6       1.8  MIN   1.7   6814     7
300    .9       2.1  MIN   2.4   7237     10
400    1.3      2.6  MIN   3     7705     13
500    1.8      3.2  MIN   3.7   8195     17
600    2.6      4.1  MIN   4.3   8676     22
700    3.8      5.4  MIN   4.9   9119     27
800    5.4      7.3  MIN   5.3   9506     33
900    7.7      9.9  MIN   5.7   9827     40
1000   11       13.6 MIN   5.9   10083    46

 PRINT WHOLE PAGE (DUMP)     (P)
 TRY OTHER VALUES            (A)
 CONTINUE:    (RETURN)        P

    extrapolations (take care!!)

 START WITH C NUMBER    >>> C  = 10
                 BIS    >>> C  = 18

INDEX K'         TMS        TZ    N(TH)   PZ
-----------------------------------------------

1000   11       13.6 MIN   5.9   10083    46
1100   15.7     18.9 MIN   6.1   10280    53
1200   22.3     26.4 MIN   6.3   10429    61
1300   31.8     37.2 MIN   6.4   10538    68
1400   45.3     52.5 MIN   6.5   10618    76
1500   64.5     1.2  HRS   6.5   10675    83
1600   91.9     1.7  HRS   6.6   10716    91
1700   130.9    2.4  HRS   6.6   10745    98
1800   186.4    3.5  HRS   6.6   10766    106

 PRINT WHOLE PAGE (DUMP)     (P)
 TRY OTHER VALUES            (A)
 CONTINUE:    (RETURN)        P
```

On critical examination of Figure (29) it will appear questionable that the N(theoretical) value should sometimes start at a low value decreasing with k, and sometimes at a high value descending with k... This is easy to understand on simply looking at a "peak width against k" regression line and comparing the "a" value with the b0 value: If a = b0, then N theoretical = n real and is independent of k, *i.e.* a constant value throughout the chromatogram of chemically similar substances.

If, however, a "bad" column is well installed, that is a>b0, or a good column is badly installed, that is a<b0, then N(theoretical) increases asymptotically with k towards n(real) or decreases with k, until all three "plate numbers" become equal. It seems hard to believe that some chromatographer colleagues have not got **this** simple message? See the literature for complaints about the ABT concept, if it interests you, but it should not. All these data are useless "house numbers", if measured with a test mixture of unknown composition, or in non-isocratic chromatography. The last-mentioned point is incredibly effective. Just polar solvent used in a non-polar column with semipolar test substances, or a high concentration of solvent in front of the later eluting test substances will ruin the possibilities of **cross correlation.** Clever selection of sample and conditions in order to make chromatography constant is the key condition for a useful system analysis. This is basic chromatography.

See the Figures and Tables.

5.8 Mobile Phase Flow Optimization

As mentioned above, each of the Tables of "full range system analysis" is only **one** point of **data** continuously changing with the mobile phase **flow**.

As also mentioned earlier, **dead time** tm is the best available measure of flow.

Let us now keep the temperature and mobile phase composition constant and change the flow.

In GC, quite a long equilibration time is necessary after changing the inlet pressure until a steady state is reestablished, *i.e.* "isocratic" chromatography. And only then are system analysis data correct.
10...20 minutes' waiting time, depending on length of column and the inlet and gas tubing volume, is normal. (Bad B-data, determination coefficients, will tell you that you have not waited long enough.)

In HPLC the equilibration time is relatively short due to the low compressibility of the mobile phase; nevertheless, some 2...5 minutes with 100 mm 3 micron packed columns is the equilibration time required under low flow rate conditions. At higher flow rates, frictional heat comes into play and the equilibration time is longer.

As GC and HPLC system analysis data are almost identical, mobile phase flow changes are documented by full sets of data but with a few HPLC experiments only and GC results—which show much more "data noise"—will only be summarized in figures.

Experimental conditions:

column: 100 mm, 4 mm i.d. steel
stationary phase: 0.3 micrometer RP C18 spherical packing
mobile phase: methanol/water 80/20
sample : homologous series of 2,4-dinitrobenzoic acid esters C3...C7
detector: UV 254 nm
pump: Latec
temperature: 20 °C
instrument: home made

We changed the flow from 0.3; 0.6; 0.9; 1.2; to 1.5 ml/min.

The results are coded under "BI—(flow in ml/min)"; "BI—0.3" means: system analysis data for this HPLC flow-experiment at 0.3 ml/min flow.

Results:

1. Chromatograms:

Please take a look at Figure (31) and (32) decide which flow is the best.
You surely will find that one cannot, or cannot easily, say which is the best flow. One cannot optimize just by deciding upon a chromatogram.
More data are needed.

2. Raw data concentrate:

2.1 a, b0, tm and statistics values

2.2 Nreal, Hreal, SNreal, Snt:

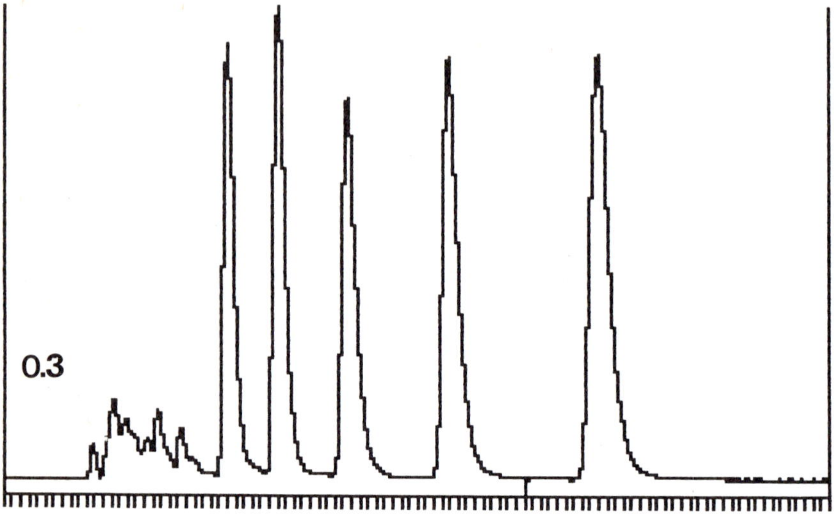

Figure (31)
Raw data of a HPL-test chromatogram run at 0.3 ml/min mobile phase flow—to be compared with Fig.(32).

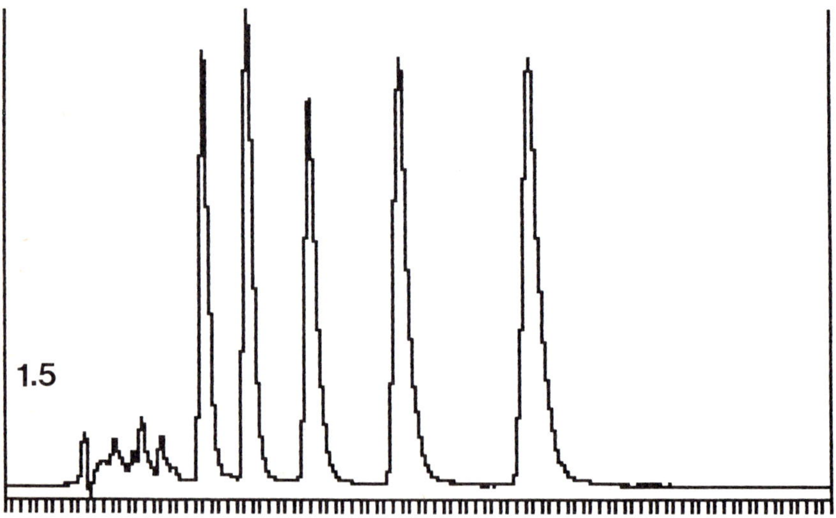

Figure (32)
As Fig.(31) but with a mobile phase flow of 1.5 ml/min. All other flow conditions were between 0.3 to 1.5 ml/min.

type	0.3	0.6	0.9	1.2	1.5
tm (s)	110.54	58.55	39.42	30.08	24.35
B for tm	0.99996	0.999995	0.999998	0.99999	0.999991
a (s)	2.71	1.39	0.85	0.63	0.56
b0 (s)	8.2 !!	3.8	2.5	2.2	1.7
Nreal	9234	9897	12003	12781	10500
Hreal (mm)	0.01	0.01	0.008	0.007	0.009
SNreal	28.7	31.4	33.2	31.6	31.1
SNt	1.41	2.91	4.58	5.72	6.96
B(b0.5/k)	0.995	0.995	0.998	0.991	0.999

Discussion:

The dead time decreases by more than a factor of four from 110 s to 25 s when the flow changes. This means that analysis is several hundred percent faster when done at higher flow. But the plate number (we prefer Nreal, which means a substance-independent measure of the "separation quality" of the column) goes through a maximum at 1.2 ml/min flow.

The separation "efficiency" (which is more important than the "quality", because it effectively quantifies how many peaks are baseline separated within the k range of zero to 10) goes through a maximum near a flow rate of 1.2 ml/min. Is 1.2 ml/min the optimum flow? Definitely **not**, as the speed of analysis (or the time necessary to accomplish a given amount of separation) increases from SNt = 4.6 through SNt = 5.7 (which is + 24% higher speed or lower time consumption) to SNt = 6.96 at the highest flow of 1.5 ml/min. This is a plus of 109% at only a minor drop in separation efficiency expressed by SNreal.

In contrast to the majority of colleagues who think in terms of optima at (or "near") the HETP/u curve minimum (see, for instance, [14], we prefer to offer the real practical working optimum: this lies between the maximum of the real separation number SNreal towards shorter dead time values, just where the falling line of SNreal over 1/tm crosses the **minimum** SN necessary for a separation to be done practically and the **minimum necessary** separation number for a given separation at the higher flow of mobile phase given highest possible separation speed, expressed in SNt (sep.number/11* tm).

On the basis of k differences in HPLC, or retention index differences in GC, one can calculate the necessary separation number exactly. And as mobile phase always will be cheaper than man- or woman-hours, the **optimum** is at the fastest possible, yet still acceptably good separation. This most practical value is determined to a considerable extent by the value of b0, the starting minimum peak width at k = zero. Therefore it is not only important to have a good column, but especially to have a good technique for installing and using the column under optimum conditions. As this continuously keeps costs at the minimum possible level, it is **the** most important factor in practical chromatography.

The following Figures concentrate numbers into graphics:

Dead time as THE measure of flow, as the much more informative quantity replacing the "velocity of mobile phase u" (in cm/s)—the latter is calculated as if the column were empty, which is not the case in the practice of HPLC—correlates excellently with the flow. So why not use 1/tm as the basis for auto-optimization of instruments?

Figure (33) gives the statistical data of correlation and linear regression.

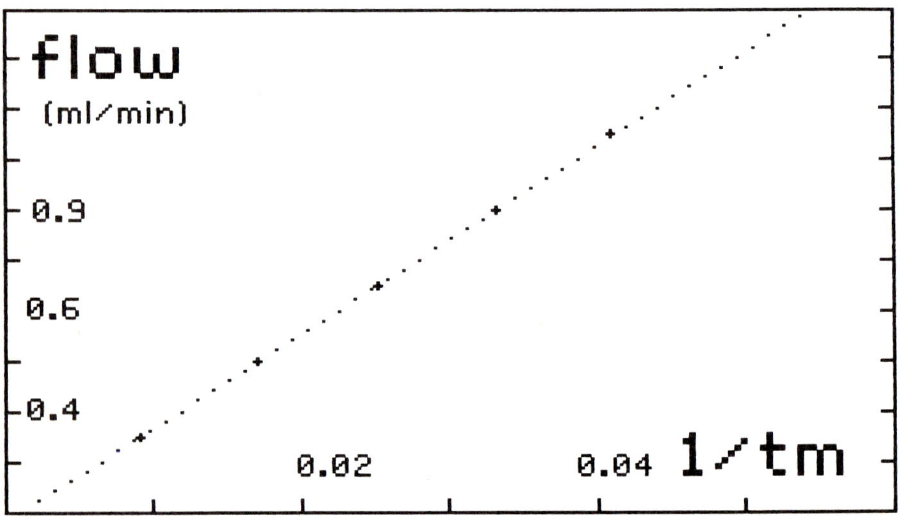

Figure (33)
Correlation between mobile phase flow in ml/min and 1/tm in sec^{-1}.

The regression quality was 0.53 ± 0.31%; B = 0.9999, see Figure (33).

When the flow changes in chromatography everything else changes too, (hopefully) with the exception of one quantity: the k values of the test substances. Cross correlation now allows checks for systematic errors in tm- or other values. One should exercise great care, however, with respect to side effects: a higher flow may change the temperature in the column packing and this will cause k data changes.

There follows an example of a check by cross correlation upon the tm calculation. It is useful to check this (or to have it checked by a cleverly written system analysis program), because in HPLC in particular the dead time iteration is not as precise as in GC.

flow	k (C3)	k (C4)	k (C5)	k (C6)	k (C7)
0.3	0.97	1.41	2.01	2.9	4.21
0.6	0.98	1.42	2.04	2.94	4.25
0.9	0.99	1.42	2.04	2.94	4.23
1.2	0.99	1.43	2.05	2.95	4.26
1.5	0.99	1.43	2.05	2.94	4.23

The k data remained constant. We can therefore use the peak-width data *versus* k as a safe measure for the separation number.

Figure (34) shows the peak width changes *versus* k as function of the flow. It gives none other than the known result of linearity over the range of chromatography used in practice. Thus decision data obtained by the ABT concept are profound.

Of course, it is questionable to use only five different flow values or the result for a test mixture of only five members of a homologous series in defining the practical optimum. But let us first continue with only five measurements and then compare our findings with some 80 data for a GC optimum.

See Figure (35) which shows a very flat SN *versus* 1/tm function, underlining the practical power of 3-micrometer packings in HPLC columns: the separation optimum is very flat. Even at the highest flow (which is the largest 1/tm value or the smallest tm) in the five runs used to find the optimum we see almost no decrease in the

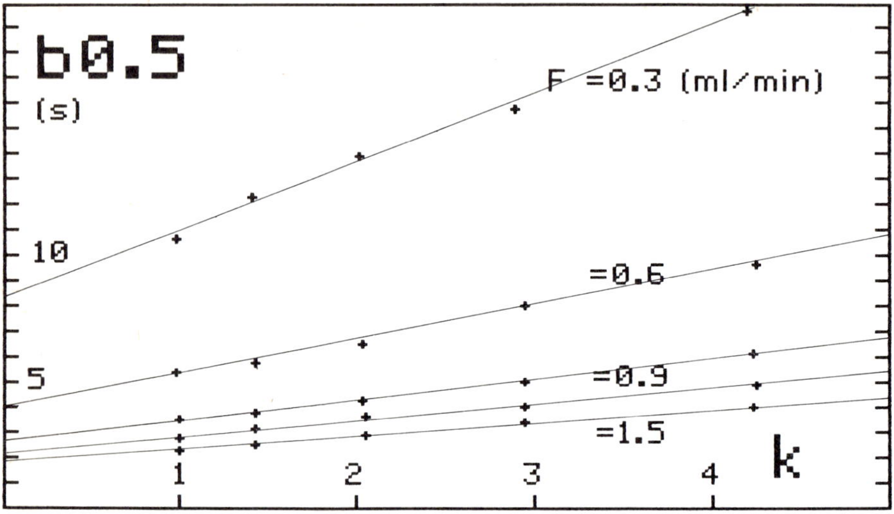

Figure (34)
Flow of mobile phase influencing the "b0.5" over "k" relation in HPLC (and GC; here: HPLC data only).

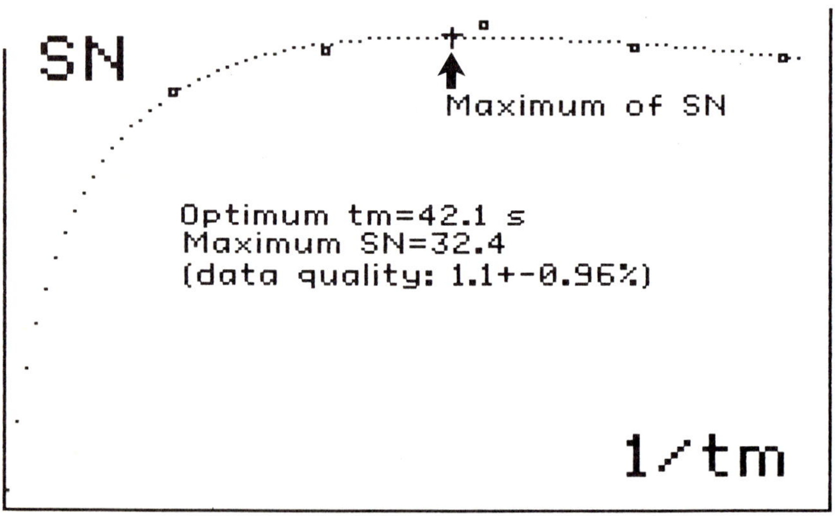

Figure (35)
Separation number over flow given as "1/tm" (as measure of the mobile phase flow or the flow velocity "u").

separation efficiency. But we do find a continuous increase in separation speed; in practice, one does not normally have pumps powerful enough to use 3-micron packed columns at the upper limits of separation speed. Or the electronics cannot follow the fast signals, or the integrators are too slow in relation to high speed HPLC runs.

What about the column packing? Does the maximum separation speed correspond to the minimum of the HETP curve? Is the maximum of n(real) at the same tm value as the maximum of SNreal?

The answer is NO. There is more than one optimum flow value. One is the optimum for maximum separation efficiency, and the other is the optimum for maximum plate numbers and in high resolution GC the plate number maximum (or plate height minimum) is found at **higher** flow than that for the maximum separation efficiency. In low resolution GC and under almost all HPLC conditions the plate number maximum is found at lower flow speed than the separation efficiency maximum. The Figure which condenses the correlating data, however, shows an inacceptably high "data noise" level. The systematic errors of the data used to calculate the plate number are not small enough.

As $n(real) = 8 * \log(2) * (tm/a)^2$, tm or "a" or both are systematically falsified. But as tm correlates well with the flow, "a", the slope of the peak width increase over k units is the inaccurate value. Let us check the determination coefficient B of the b0.5/k regression: B is —see above—only in the range of 0.99 to 0.999. This is NOT precise enough.

One cannot measure time and peak width precisely enough (the reader might remember this repeated statement).

Or we must use MORE than just 5 differing flows, or we must repeat analysis at an adjusted flow. The latter is the best solution of the problem, because if there is a data drift from run to run it tells us something about the stability as well as the equilibrium reached on changing the flow.

Figure (36) underlines this and Figure (37) is the practical answer.

Conclusion

Flow optimization is not easy. The data necessary to find the real optimum are prone to systematic errors. Repeated measurements and critical statistical tests are a must.

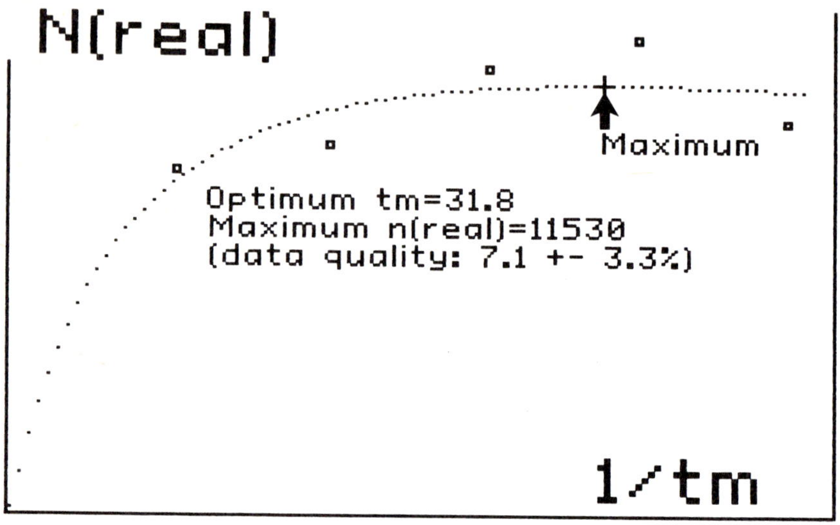

Figure (36)
Theoretical plate height over flow expressed in "1/tm".

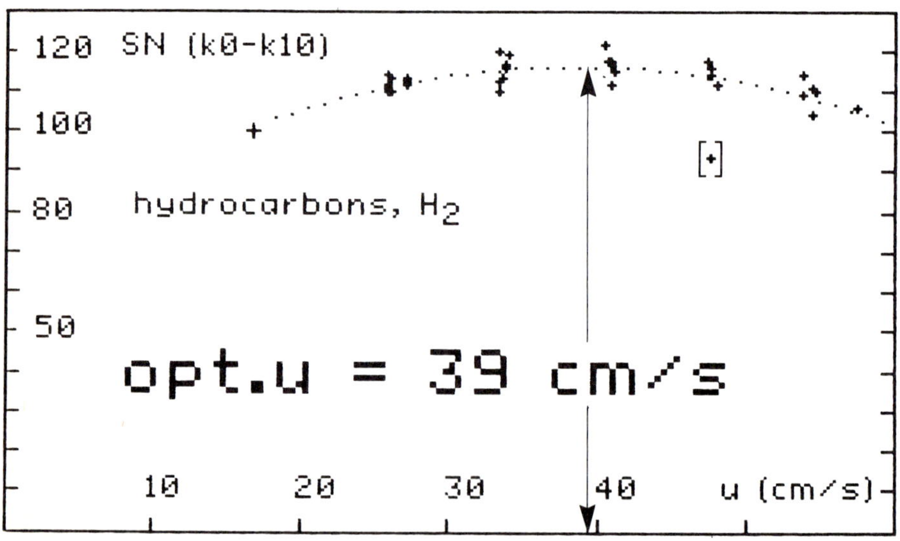

Figure (37)
Mobile phase flow optimization in capillary chromatography. The data are in contrast to theoretical values, but not excessively so. An optimum is found only by many more measurements than used normally. Here over 40 full system analyses have been combined to one final result.

Here is an often repeated measurement for capillary GC data, where SN maxima and h(real) minima are not easy to find due to the utmost data precision necessary in high resolution chromatography. See Figure (37).

Experimental conditions pertinent to the data in Figure (37): the same as used throughout the Sections of this book dealing with computer GC.

The majority of the measurements have been concentrated automatically by a BASIC program which reads T-files automatically, analyzes the contents, and writes data concentrates in appending files. Thus a huge amount of experimental data can soon be summarized and translated into graphics. Although the graphics resolution of the system we used (normal Apple II graphics) is far too low, graphics can be read by humans at a rate of 1 to 10 mio bit/s but numbers can be read only at the low level of 12 to 16 bit/s. This makes even poor graphics superior to number-based decisions. The precision ultimately needed is called from the stored values.

In closing this section on system analysis, which is based upon peak width at half height data instead of sigma measurements at 60.7% height of a "Gaussian" peak, we should check some data of real Gaussian peaks.

It is simple to construct precise Gaussian peaks. So system analysis algorithms can be checked with real-life chromatography peaks *versus* Gaussian peaks. Although more and more chromatographers **know** that peaks cannot be Gaussian shaped (neither theoretically nor in practice) even some 1982 papers are written on the basis of the Gaussian shape error.

We analyzed perfect Gaussian shaped peaks. We used only 11 raw data over the range of peak width at half height. We used an algorithm which calculated "raw data" from a peak of 10.000 units of theoretical peak width at half height and 10000 area units.

Peak shape analysis results are shown in Figure (38).

The "sigma" value is perfectly in line with the shape of the curve.

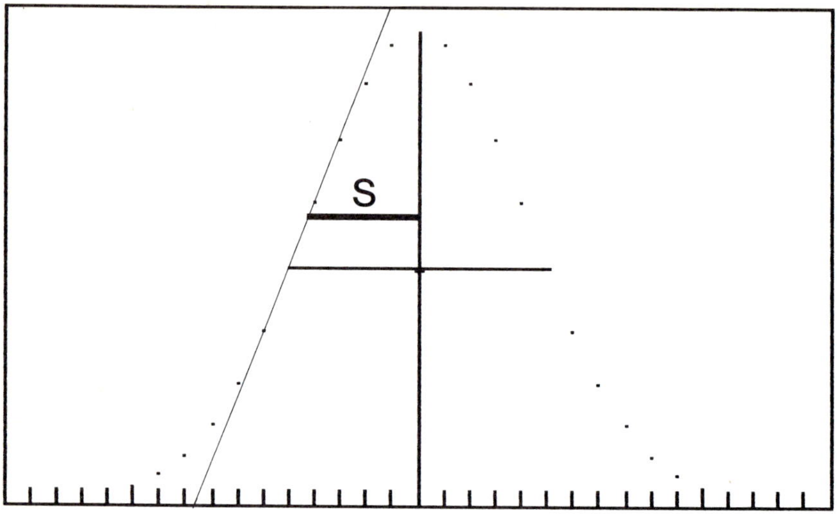

Figure (38)
A perfect Gaussian shaped peak (calculated) is analyzed by the shape analysis program. "sigma" (or "s" at 60.7% of peak height) is perfect, just touching the signal shape curve (seen to touch the tangent on the left-hand side of the peak).

	measured		theory	error (%)
Peak width at half height:	10.099	instead of	10.000	0.99%
Area:	9998	instead of	10,000	−.02%
"Sigma" (= s):	4.242	instead of	4.2466	−.11%
Variance:	17.993	instead of	18.03	−.21%

This appears good enough as an absolute error, because most of the calculations in chromatography are relative data and absolute errors are not too disturbing, at least when they lie in the .1% range.

So in fact you can trust raw data digitization into only about 10 values per peak width range at half height. Results based upon this rather "diluted" data density are mathematically in good shape and practically in good agreement with the time axis accuracy needed for acceptable system analysis.

6 Computer Supported Chromatographic Analysis

Qualitative chromatography (GC as well as HPLC) is widely underdeveloped. Only a few laboratories make use of k values in GC and HPLC and retention index data in GC. It is this which can be supported elegantly by on- or off-line lab computation and the results are of great help for quantitative analysis.

Automatic calculation of k- and index data is not without problems as both data groups need dead time values and therefore isocratic runs. Most practitioners work under non-isocratic conditions in GC and non-isocratic conditions in HPLC but only in routine analyses. Chapter 6 discusses computer support for retention index and k values based upon the raw data sets used throughout the volume. Emphasis is on inter- and extrapolation of the data, as this is beneficial in practice.

In the case of TEMPERATURE PROGRAMMED gas chromatography non-linear regression is a must for retention index calculation. Detailed data sets are given and the BASIC PROGRAM for 1st...6th order polynomial regression is given in full length as source code in the appendix.

As quantitative analysis is handled well in most cases by on-line computerized electronic integrators, this volume deals only with VIDEO INTEGRATION as an aid for otherwise hopeless tasks: quantitation one order of magnitude below the levels possible up to now when NO computer is available. Thin-layer chromatography is improved towards high performance TLC when done on-line (automatically) by a lab computer and the "structure" under the peak signal is deducted.

Concentrate:

Retention index in isothermal and temperature programmed gas chromatography and k in GC and HPLC support quantitatively accurate analyses. Non-linear regression is a must for qualitative analysis by TGC. VIDEO INTEGRATION is the procedure for quantitation in GC, HPLC, and especially TLC, changing the latter into HPTLC which is now as accurate as GC and HPLC but much faster and less expensive in mass data analysis.

6.1 Qualitative Chromatography Analysis Supported by Computer Chromatography

In HPLC there is no consensus between the majority of users about a qualitative analysis system. k = (tms − tm)/tm based upon k reference data (like those we use for HPLC system analysis) would be a possibility. But the concept of a dead time, which would be essential for working with k data calculations, is not widely accepted. The procedure adopted will very probably be multiwavelength simultaneous analysis, performed by on-line computer. Intelligent data reduction will remain a problem for a longer time.

This multi-simultaneous spectral analysis of the eluate will be discussed in one of the forthcoming volumes of "Computer Chromatography", where simultaneous double or multi-column and double or multi-detection analysis will be treated.

We will concentrate here on retention index analysis in gas chromatography.

Retention index calculation is now possible with some of the computing integrators on the market.

The retention index is a relative-over-relative value and is thus especially insensitive to systematic identification errors, provided that it is correctly determined. It is incorrect to use constant relations of qualitative calibration substances and it is "bad laboratory practice" to use normal alkanes only as standard substances.

The retention index in isocratic (isothermal) GC is defined as the logarithmically **interpolated** position of a peak relative to the position of immediately adjacent neighboring qualitative standards of known retention index. Well: the historical definition of the authors *Wehrli* and *Kovats* [15] differs from the information written here, but practice has shown that the index system is of practical utility only when the qualitative calibration standards actually enclose the peak position of a substance to be identified and when the standards are of a similar chemical nature.

Example:

If the mixture to be analyzed belongs to the class of unsaturated alcohols, then some unsaturated standard alcohols should be used as retention index standards.

In order to keep an internationally acceptable identification standard **system** available to each chromatographer, normal alkanes were chosen by *Kovats* and *Wehrli* as standard substances. They are readily accessible, stable, and limited to about 30...50 compounds only of wide distribution. In natural gas from C1 to C4, in gasoline from C5 to C12, in diesel fuel from C10 to C32, in candles from C20 to C40...

The BASIC algorithm for retention index calculation is

I = (In − Iv)*(log((tmsx − tm)/(tmsv − tm))/log((tmsn − tm)/(tmsv − tm)) + Iv

Iv = index of a standard substance eluted before the substance x to be identified.

In = index of a standard eluted behind x
tmsx, −v, −n are the non-adjusted retention times of substance x, v, n.

The algorithm for temperature programed retention index calculation is

$$Ix = A0 + A1*tms + A2*tms^2 + A3*tms^3 + A4*tms^4 + A5*tms^5$$

where the factors A0,...A5 have been found by fifth order polynomial regression of the nonadjusted retention times of seven standard compounds enclosing all of the substances to be identified.

This algorithm does not agree with those of many authors, especially of the inventors of the retention index system. However, it is of remarkable practical utility. Up to now, all retention index data calculated by this 5th order polynomial based upon seven standards resulted in a value which was—within the range of experimental error—**identical** to the isothermal retention index measured at the elution temperature of properly performed temperature-programmed gas chromatography.

A wider range of experimental conditions, like

starting temperature
heating rate
end temperature

showed only a minor influence on our 5th order polynomial retention index value.

The retention index of *n*-alkanes is known under all conditions of stationary phase, temperature, operating mode, phase concentration, type of carrier material, mobile phase, column length, inlet/outlet pressure ...*etc*.

In any case, the retention index equals 100 x C atom number of the *n*-alkane. There is no problem encountered in using the retention index in all cases with all detectors. Electron capture detectors detect *n*-alkanes sensitively enough when used at elevated temperatures. Any detector can be used parallel to a flame ionization detector, in which *n*-alkanes are detected sensitively enough.

If *n*-alkanes have once been used together with homologous series mixtures of specific high detectability, like fluorine-containing fatty acid esters (trifluoroacetic esters, for instance) then identification based upon enclosing standards of known retention index is very practical.

If the **adjusted** retention time ts = tms − tm is measured precisely enough on the 10 millisecond level using polynomial interpolation and if the temperature is then kept constant under isocratic conditions, the resulting retention index data are (for some chromatographers incredibly) reproducible and often also accurate. In cases of *n*-alkane enclosing standards, the accuracy might be in the range of below one index unit, often down to 0.1 unit, even as interlaboratory repeatability. Well done index calculations can reach the 0.01 or even better intralaboratory repeatability level when utmost care is taken and everything has been done to avoid systematic peak maximum shift effects. (Oversampling, using the wrong solvent in the wrong chromatographic position, *etc*.)

But more details are not to be discussed here. The danger of systematically falsified, otherwise extremely well reproducible retention index data has been mentioned enough in the additional remarks.

There are two key conditions, which demand attention in practice:

retention index data cannot be accurate if calculated with a wrong dead time and interpolated in a wider range than 100 units. Therefore the following algorithms have been written taking into account this condition of the limited range for interpolation: 100 units only.

A dead time **is wrong,** if it does not allow a linear regression of log(ts) over standard substance indices. The methane dead time **is** wrong. It makes the interpolation line in between enclosing standards "hang like an overland telegraph wire".

But the calculation is done linearly according to the triangle rule and nobody sees the error (in simple ways), because only the enclosing standard substances are known by index.

A cleverly written retention index program therefore

- calculates tm
- checks for log(ts)/index linearity (B = >0.99999!)
- checks for neighboring and enclosing standard substances
- checks for an interpolation range of 100 or below 100 units (in case of non *n*-alkanes as additional standard substances)
- warns the user when extrapolation instead of interpolation happens
- and allows the calculation of non-adjusted retention times from index data as well, for **cross correlation**.

Table (11) gives data examples. The index calculation algorithms which take into account the 100 units range limitation as well as the "back calculation index" into tms data are given elsewhere.

Data for Table (11):

The index calculating program found 4 consecutive *n*-alkane standards (I = 800....1100) tm could be calculated by procedure 2 (p. 65) and resulted in

$$\begin{aligned} tm &= 234.155 \pm 0.583 \\ B &= 0.9999991 \end{aligned}$$

tm could also be calculated by iteration using the linear regression log(ts) over Index:

$$\begin{aligned} tm &= 232.69 \\ B &= 0.9999997 \end{aligned}$$

As the second tm value has the better determination coefficient it was selected for the (automatic) index calculation of all other tms data in the data file; see Table of raw data, which has been enlarged to a full documentation of a "retention index session".

Note: Only in isothermal GC are retention indices reproducible to such a level that not only identification but even structure elucidation is successful. Using only two differing stationary phases (SE30, OV17), *Post* [16] could positively identify about 600 compounds of toxicological importance using a synoptic nomogram.

We will discuss the simultaneous use of more than one stationary phase for purity tests and positive identification in Volume 2.

Under temperature programmed GC conditions retention index calculations do not need dead time data, but non-linear interpolation, as mentioned above. This will now be shown in a data example.

See Figure (39).

6.2 Temperature Programmed GC for Qualitative Analysis

Temperature programming is widely used in practical GC, but is not all too easy a technique. 15 years ago commercially available GC instruments could reproduce TGC retention time data to the good level of 1 second in 1000-second long runs, or to 1-mm peak position accuracy on 1 m-long chromatograms. This was possible by working strictly with a clock: Start of the temperature program, moment of injection, hold time, cooling time and restart of the temperature program were controlled in cyclic manner by the second. After the third or fourth full working cycle the instrument was stable. The equipment has not improved all that much in the last 15 years. And there is no help forthcoming from any computer program if heat memory, fluctuating mains frequency or voltage, varying working cycle are dominant and sharp identifications are wanted.

```
        **  CALCULATION OF INDICES GC  **

                   --IFC--


 PROGRAM LIMITS DATA INPUT TO
 100 VALUES

 DATA INPUT FROM DISK ?                (Y:N)
                                          Y
CALCULATION BASED ON N ALKANES     (A:N)?A
 (?) = CATALOG
 FILE NAME : C.INDEXDA1.1

NR          TMS        INDEX      B0.5
-----------------------------------------
FILE CONTAINS = 13 DATA

1           365.3      800        0
2           372.2      0          0
3           388.9      0          0
4           407.6      0          0
5           463        0          0
6           470.7      0          0
7           485        900        0
8           489.9      0          0
9           700.1      0          0
10          713.9      1000       0
11          789        0          0
12          1001.8     0          0
13          1149.5     1100       0

'RESULTS'BASED'UPON'NmALKANE'PROCEDURE'


>>       MEAN OF TM       = 234.155
STANDARD DEVIATION        = .503
                    IN % = .249

FOLLOWING DATA USED FOR TM ITERATION


     TMS        INDEX
 ----------------------------------------
     365.3      800
     485        900
     713.9      1000
     1149.5     1100
      TM BY ITERATION (IFC)
PLEASE WAIT
                 B= .800842
                                  TM = 364.934
```

```
                    B= .999667
                              TM = 211.508

                    B= .999762
                              TM = 215.161



                    B= .999999
                              TM = 233.061

                    B= .999999
                              TM = 232.696


 RESULTS

*TM BY   ITERATION            = 232.69 (S)
                            B = .999999714

*TM BY 'N ALKANE PROCEDURE = 234.15 (S)
                         +-S = .249 %

*DIFFERENCE OF BOTH TM'S    = .62 %


ALL FILE DATA INTO INDICES              (D)

DEAD TIME 'TM2' BY ITERATION ?          (T)

END PROGRAM  ?                          (E)
                                         D

   **  CALCULATION OF INDICES GC  **

               --IFC--


 PROGRAM LIMITS DATA INPUT TO
 100 VALUES

 DATA INPUT FROM DISK ?                (Y:N)
                                         N
CALCULATION BASED ON N ALKANES   (A:N)?M
   **  CALCULATION OF INDICES GC  **

                 IFC


YOU NEED PRECISE TMS DATA FROM AT LEAST
FOUR CONSECUTIVE N ALKANE HOMOLOGUES
```

```
ENTER C NUMBER OF THE SMALLEST
THEN OF THE LARGEST HOMOLOGUE

ENTER TMS DATA IN ASCENDING ORDER

* ENTER C NUMBER OF THE

  ENCLOSING HOMOLOGUES  :


C MIN = 8
C MAX = 11

START WITH SMALLEST TMS VALUES:

     TMS = 456
                    FOR  C8

     TMS = 671.6
                    FOR  C9

     TMS = 1083
                    FOR  C10

     TMS = 1872.3
                    FOR  C11


'RESULTS'BASED'UPON'NmALKANE'PROCEDURE'


>>      MEAN OF TM       = 221.3
STANDARD DEVIATION       = 1.17
                   IN % = .528

    (4 N ALKANES)

    COEFF. OF DETERMINATION

                       B = .999999421

ALL FILE DATA INTO INDICES              (D)

DEAD TIME 'TM2' BY ITERATION ?         (T)

END PROGRAM  ?                          (E)
                                         D

WHICH TM VALUE : (1)=232.6 (2)=234.1
                   2
1    K'= .57     T= 365.3     I= 800
2    K'= .6      T= 372.2     I= 807.886
```

```
3     K'= .67      T= 388.9      I= 825.463
4     K'= .75      T= 407.6      I= 843.041
5     K'= .99      T= 463        I= 885.817
6     K'= 1.02     T= 470.7      I= 890.93
7     K'= 1.08     T= 485        I= 900
8     K'= 1.11     T= 489.9      I= 902.979
9     K'= 2.01     T= 700.1      I= 995.493
10    K'= 2.07     T= 713.9      I= 1000
11    K'= 2.39     T= 789        I= 1022.498
12    K'= 3.31     T= 1001.8     I= 1072.748
13    K'= 3.94     T= 1149.5


WHICH TM VALUE : (1)=232.6 (2)=234.1
                    1
1     K'= .56      T= 365.3      I= 800
2     K'= .59      T= 372.2      I= 807.906
3     K'= .66      T= 388.9      I= 825.515
4     K'= .74      T= 407.6      I= 843.106
5     K'= .98      T= 463        I= 885.846
6     K'= 1.01     T= 470.7      I= 890.949
7     K'= 1.07     T= 485        I= 900
8     K'= 1.09     T= 489.9      I= 902.984
9     K'= 1.99     T= 700.1      I= 995.499
10    K'= 2.05     T= 713.9      I= 1000
11    K'= 2.37     T= 789        I= 1022.511
12    K'= 3.28     T= 1001.8     I= 1072.761
13    K'= 3.91     T= 1149.5

TMS INTO INDEX OR INDEX INTO

TMS ?                  (E=END)    (T:E) ?T

BERECHNUNG RICHTIGER INDEXWERTE

RESTART BY (0); INDEX IN TMS: (-1)


TMS = 671.6

 K'= 2.01    I= 900            TM = 222.8

TMS = 671.7

 K'= 2.02    I= 900.034        TM = 222.8

TMS = 671.8

 K'= 2.02    I= 900.068        TM = 222.8
```

```
TMS = 1872

 K'= 7.4    I= 1099.972       TM = 222.8

TMS = 1872.3

 K'= 7.4    I= 1100           TM = 222.8

TMS = 22400
     OUTSIDE CALIBRATED
             RANGE. UNSAFE !

 K'= 99.55   I= 1499.139     TM = 222.8

TMS = -1
SWITCHED TO INDEX --> TMS CALCULATION
INDEX = 100

TMS = 225.2  TM = 222.7

** EXTRAPOLATION IS UNSAFE !!
SWITCHED TO INDEX --> TMS CALCULATION
INDEX = 200

TMS = 227.4  TM = 222.7

** EXTRAPOLATION IS UNSAFE !!
SWITCHED TO INDEX --> TMS CALCULATION
INDEX = 300

TMS = 231.7  TM = 222.7

** EXTRAPOLATION IS UNSAFE !!
SWITCHED TO INDEX --> TMS CALCULATION
INDEX = 400
```

Table (11)
A full "retention index session" starting with data input (from a disk file—see first Table of raw data with the file name "C.INDEXDA1.1") to the "manual calculation" of index data.

If everything is kept reproducible, however, then we find 5th order polynomial interpolation of the retention time or retention temperature over the retention index of standard substances to be a good basis for precise identification and quite acceptable retention index data. In TGC about 8 to 12 homologues can be separated within one run.

We used a capillary of 14000 mm length, and coated with a thin film of medium polarity type, and worked with a temperature program starting at 40 °C with a rise of 2 ° min^{-1}.

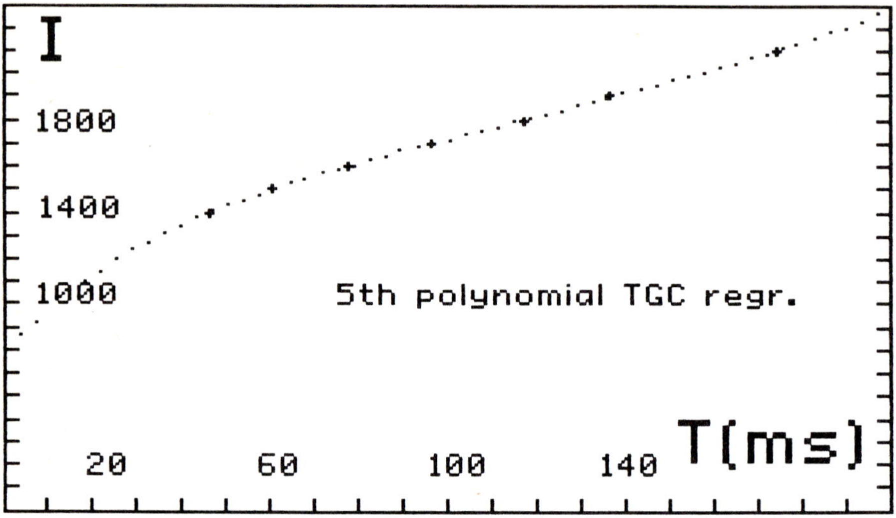

Figure (39)
5th order polynomial regression through 7 raw data pairs "retention index over retention TEMPERATURE tms in °C" are basis for retention index calculation in temp.progr. GC. Seven consecutive n-alkanes have been used as calibration standards.

There follow some data given as retention temperature (the temperature corresponding to the peak maximum elution time tms) over retention index of *n*-alkanes C13 to C20.

no.	T(ms) °C	retention index
1	46.2	1300
2	60.3	1400
3	77.7	1500
4	96.5	1600
5	116.9	1700
6	136.9	1800
7	174.4	2000 (not 1900!)

We used 5th order polynomial regression to interpolate the function.

The statistical results were

determination coefficient	B = 0.999993
standard error	1.547
regression quality	0.03 ± 0.02%

polynomial factors

A0 = 648.84 A1 = 23.8688 A2 = − 0.3091
A3 = 2.575E-3 A4 = − 1.091E-5 A5 = 1.874E-8

and as the index values are quite large, the factor A5∗(index^5) is in the range of several units. This justifies 5th order polynomial interpolation.

If we were to use only 1st, 2nd, 3rd, or 4th order regression the statistical data would tell us there is a bad fit:

Same data as given above result in

order	B	regression quality
1	0.998	0.57 ± 0.47%
2	0.999	0.46 ± 0.19%
3	0.9999	0.08 ± 0.06%
4	0.99999	0.03 ± 0.03%
5	0.999993	0.03 ± 0.04%

One will doubt that there is any real improvement when comparing index data calculated on the basis of 2nd...4th with those obtained by 5th order polynomial interpolation.

There follow some results for a substance eluting at T(ms) = 75 and one eluting at T(ms) = 165 °C

Index	2	3	4	5 degree
	1476.46	1485.54	1486.58	1486.12
	1949.54	1945.45	1948.77	1944.91

Many users of gas chromatography interpolate **linearly** in the temperature programmed mode.

In our data example this would result in
1427 for the first and
2003 for the second compound
of the data example.

In cooperation with some colleagues at a computer chromatography training course, we have now analyzed the two substances at 75 and 165 °C constant GC column temperature: no agreement was found with the linearly interpolated data but good agreement was observed with the data interpolated by the highest order polynomial.

In HPLC, log(k) shows quite a good linear dependence on the composition of mobile phase mixtures, using binary systems like water/methanol, acetonitrile/water, *etc.,* mixtures.

Real-life HPLC does not work all that well with binary mixtures for the optimal mobile phase composition. The third and sometimes the fourth solvent added in lower concentration "does the trick". log(k) no longer shows too simple a dependence on the mobile phase composition and so computer use in qualitative HPLC becomes unavoidable. This is true for simultaneous multi-wavelength detection as well as binary/ternary/quaternary mobile phase composition and the peak position in the chromatogram.

See the additional literature at the end of this volume.

6.3 Quantitation in GC and HPLC

Most chromatographers believe in linear detector calibration curves but are aware of the often limited range of this linearity. We have tested many detectors by the exponential dilution calibrating procedure. This is the only satisfactory working calibration procedure if a working range of more than one order of magnitude has to be checked.

Let us first believe in linearity.

In this—and **only** in this—case the following fundamental formulas hold true:

$$\frac{\text{area peak A} * \text{fA}}{\text{area peak B} * 1.00} = \frac{\text{weight A (gr)}}{\text{weight B (gr)}}$$

area = peak height times peak width at half height, or the integral as measured by the integrator

A = analyte

B = internal standard, external standard, or any other calibration substance.

fA is the substance-specific correction factor, which allows us to end up with a quantitative result in weight-% concentration. The best answer for quantitative analysis questions, as it will be understood by most of the partners of analysts is "weight-%".

As concentration (A) = 100 * gr (A) found/sample used, we simply have to replace gr (A) and gr (B) in the last formula by areas, the correction factor, and internal standard % values and end up with the well known approximative formula for quantitative chromatography:

% A = (%B * area A * fA)/(area B)

and on replacing "area" with "integral" we only have to guarantee

a) that the analytical detector signal correlates linearly with the amount of substance and

b) that the integrator integrates linearly.

Within limits—which are often too narrow to believe—this basic quantitative formula can be used for quantitation.

"Area" is easily written but not easily integrated, as a considerable part of the signal is often unknown: this unknown component of the signal may correspond to the so-called "baseline" where all the small signal-over-time changes are located and it might be a special "separation" line, *i.e.* the boundary between overlapping peaks.

Any chemist who has worked in trace analysis will know that large quantities of many "traces" are often located "under" a main peak.

As **no** information whatsoever can be gained under such circumstances, no simple integrator can accurately perform quantitation. Data which show excellent reproducibility are not necessarily correct.

These problems have been known and the subject of discussion for some 20 years. The frequently heard advice to "separate better with a good column, the computer cannot do it" is only partly true. If the computer has more information—for instance from a second detector channel—and if the integration program based upon cross correlation works, then the computer can help very much.

This double- or multi-channel based quantitation support is treated in subsequent volumes.

The next source of computer aid in quantitation is the **linearization** of the detector signal. It is perfectly easy to feed a computer with an experimental calibration function measured by an exponential dilution calibration. These data permit continuous linearization. We will discuss this important practical step in one of the forthcoming volumes.

We will conclude the present Volume One with some remarks and a few examples of "Video Integration".

6.4 Video Integration

There are some analytical conditions, especially in trace analysis and in the analysis of highly complex samples like "motor oil", butter, wine extracts, or the "non-volatiles" of crude oil distillate...a seemingly endless list, where electronic integrators and even computer programmable integrators fail. In such cases the skill of the operator is the only chance to solve the problem or at least to get some results for a particular value.

It is a fact that any technical skill can be translated into computer programs, but costs may be prohibitive. In such cases, a critical look at the screen showing the critical chromatogram might help. Decisions where to position the baseline and from where to where integration is desired can be done by positioning "cursors".

In the case of "critical" or bad chromatograms, the degree of consensus between chromatographers about what to do and where to do it is very limited. This will also be apparent in the comments passed on the figures shown here.

All we want to show are the possibilities of combining special human skills in critical questions of quantitation at the (mathematically and statistically) permissible limits with computer help.

Perfect repeatability is ensured by "Video Integration".

This is easy to understand:

We are talking about a man-machine combination in deciding what is the baseline and what are significant results in a signal, which can be integrated completely after deduction of the baseline under the peaks.

The BASIC program is then simply:

```
FOR P = 1 TO END

REM "DECIDE ON PEAK BEGIN & PEAK END
NUMBER THE PEAKS CONSECUTIVELY"

P = P + 1
NEXT

FOR J = 1 TO P
FOR I = 1 TO N
A = A + H(I)

REM "H(I) IS THE RAW DATA SIGNAL CORRECTED BY
THE BASE LINE HEIGHT POSITION"
```

```
NEXT
PRINT J, A(J)
REM "A(J) IS THE INTEGRAL OF PEAK J"
NEXT
```

See the Figures (40), (41), and (42) and comments, which should give a clear picture about possibilities and limitations of video screen supported integration in non-routine complicated or non-automatically integratable chromatograms.

In order to dispel any remaining doubts about whether a computer really can help to improve (and that dramatically) the analytical situation beyond that existing today, with its perfect hardware programmed electronic computer integrators, we close this volume with a "chromatogram".

This "chromatogram" is at first absolutely untreatable. It is practically impossible to integrate anything. A "hopeless" situation? This "chromatogram" can be repeated in all aspects of its "structure" to within at least ±2...4% by quantitizing a thin-layer plate with an optical scanner. The structure is caused by the layer particles and light reflectance.

Thin-layer chromatography is an analytical field which has received very little mention in this Volume 1 of "Computer Chromatography". However, it is of tremendous importance as TLC, well done, is the fastest, least expensive, and **most accurate** quantitative chromatographic technique known to date. The possibility of unlimited detection procedures and truly simultaneous multi-sample analysis is of enormous analytical value.

But look at the "chromatogram" Figure (40)! The Figure immediately shows why many analysts know nothing about TLC as a technique of quantitation. How can this procedure produce "accurate" results?

The explanation is to be found in the possibility of performing simultaneous analysis plus multi-calibration in a single, uniform chromatographic systems—as uniform as a solid/liquid mixed bed system can be.

Our goal was to further improve TLC separations to one order of magnitude below the lowest detectability limit as defined up to now. We stored the raw data of this "chromatogram" (2000 measurements per 40 mm chromatogram or "track" length, corresponding to 50

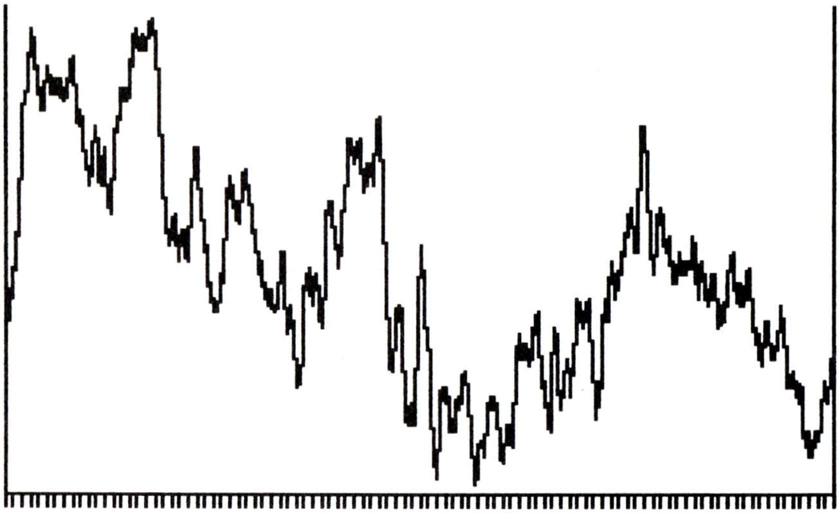

Figure (40)
A "hopeless" chromatogram. TLC below the detectability limit as defined by classical TLC. Well reproducible "chromatogram", but not translatable into any useful numbers.

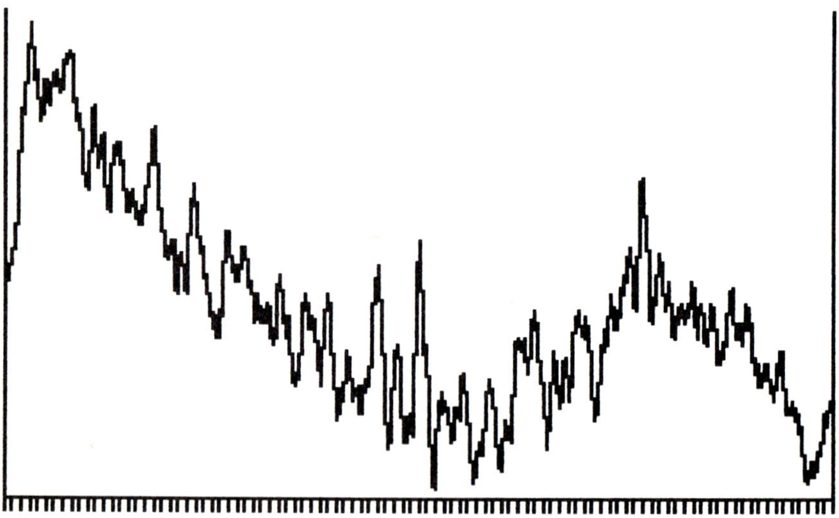

Figure (41)
"Baseline" under the chromatogram of Fig.(40). Well reproducible. Reason for this "baseline": structure (particle distribution) on the thin layer.

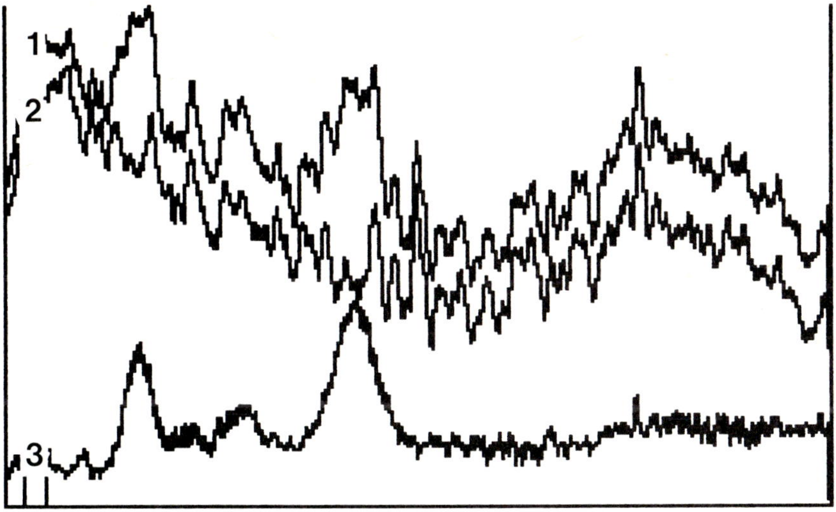

Figure (42)
Line 1 = Fig.(40). Line 2 = Fig.(41). Line 3 = difference signal minus structure, which is a thin layer chromatogram, now well above detectability limits.

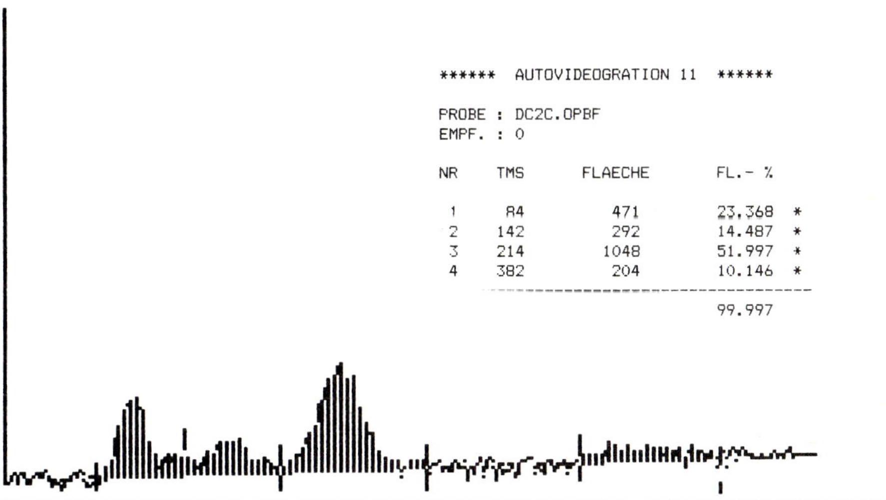

Figure (43)
Integration of the difference of raw signal minus structure (line 3 in Fig.(42)) by "AUTO VIDEOGRATION", a program which combines baseline correction, and manual and automatic integration in any combination. Raw data are loaded as binary files at high speed. A total of 1000 raw data was used throughout Fig.(40)...(43).

measurements per mm or one measurement every 20 micrometers). But **before** using the TLC plate for a separation we washed it until it was stable and quantitized the "empty" plate, thus acquiring data for a real baseline.

Now—at least at first glance—things became simple:

These are the raw data (see Figure (41)):

```
REM "M = BASELINE RAW DATA MEASURE"

FOR I = 1 TO 2000
B(I) = M
NEXT
```

At a constant A/D sampling rate and a constant speed of plate movement under the scanner light beam, the raw data counter "I" represents the position of "particles" on the plate and thus permits recording of a "structured" repeatable baseline signal.

This was—at first glance—easy to deduct from the final analytical signal:

```
REM "A = ANALYTICAL SIGNAL FROM PLATE
SCANNER"

FOR I = 1 TO 2000
S(I) = A
NEXT

REM "K(I) = CORRECTED BASE LINE AND STRUC-
TURE FREE CHROMATOGRAM SIGNAL"

FOR I = 1 TO 2000
K(I) = B(I) — A(I)
NEXT
```

In practice, it is impossible to position the plate with an accuracy of 0.02 mm.

The slightest change in the speed of plate movement, or in the timing—0.1% change was enough—resulted in data that were less correct than those measured without any correction!

To achieve at least a partial solution of the problem we developed, together with *A. Miller* [17], a program for optimization of the correction.

A short BASIC program changed the raw data "counter" and checked the standard deviation of the differences

"K(i) = B(i) − A(i)" until a "counter correction" was found which corrected the mechanically induced inaccuracy of the plate positioning.

The ensuing Figure (42) show the effects and the final result.

Using "Video Integration" we could quantitize the "signal difference" TLC minus "plate structure" to a repeatability level of plus or minus 2...3%.

Most of the existing electronic integrators are fooled by signal structure and automatically lower the peak detectability limits. Thus large quantities of peak area representing substances are cut off as baseline.
We fed the digital signal, corrected after deduction of the structure and after reconversion from digital to analog form into the integrators and ended up with most astonishing improvements. Now only real noise—easy to be handled by electronic integrators—had to be suppressed.

In fact, this way of integrating with electronic integrators having high speed optimized integration algorithms is better than direct on-line integration (see Fig.(43)):

- digitize the chromatogram signal
- store the raw data
- treat them statistically, or systematically—by differences, linearization, *etc., etc.*
- reconvert from digital to analog form
- send the improved analytical signal under the best conditions for the integrator INTO the integrator

Nonsense?

Try it. You will change your mind. The first chromatographic analysis of an important sample you have to repeat will pay back your additional costs. If not, then the second one will. As a chromatographer, I have had to redo a lot of injections, or secondary sampling. More expensive was primary sampling and particularly serious was the position of the analyst in the case of **no** further sample material, or fast changes, or unique situations.

Raw data are of the utmost importance. Ready access to today's computers and the possibilities of treating raw data while still preserving the original raw data set makes chromatography easier and ever more powerful.

Appendix

Hardware and Software Used

Hardware, Chromatography

Gas chromatographs:

Carlo Erba (Italy) Fractovap series 2350, Electrometer 180
DANI (Italy) model 3900
Becker Packard (The Netherlands) model 419 (10 years old!)
Intersmat (France) model IGC 120 DFL

Liquid chromatograph:

Home made (IfC) but with the Spectra Physics (USA) pump model SP 8700

High performance thin-layer chromatography:

CAMAG (Switzerland) TLC/HPTLC scanner, anticircular separation chamber, anticircular/circular scanning mode

Samples:

n-alkanes C6...C10 in C11 as solvent
homologous esters in the last member as solvent
2,4-dinitrobenzoic esters C3...C8 in the mobile phase as solvent

Columns, capillaries, plates:

GC: fused silica 16,000 mm length, 0.24 mm i.d., SE30, produced by Antech GmbH

HPLC: 3 micron spherical RP silica ODS, in 100 mm x 4 mm thick walled stainless steel tube from K. Bischoff—Stuttgart, W-Germany

HPTLC: Nanoplates Si60 "HPTLC" Fertigplatten Kieselgel 60 F254" from E. Merck—Darmstadt, W-Germany

Hardware, Computer

Apple II plus (Euro and US-version)
incl. language card, 48kRAM, disk drive, Epson printer MX-80 FT plus Graphtrax and MX-82 FT; Sanyo video monitor

IfC-Chromputer ® from Antech GmbH Bad Duerkheim, W-Germany with 192 kRAM and 21 megabyte ICE hard disk from ICE Ashford, Middlesex, UK

12 bit A/D & D/A converter "Adalab" from Interactive Microware Inc., State College, PA, USA

Software

From Interactive Microware (see above):

QUICKI/O, Curfit & Sciplot from P.K.Warme, Interactive Microware Inc., USA

Own software from Institute for Chromatography, Bad Duerkheim, W-Germany:
Polynomial Regression "POLYN";
Auto Videogration
Auto System and Shape Analysis
DAC (digital-to-analog replotter for Chromatography)
Index General for Chromatography
Raw Data and Measurement Package
VIN Videointegration
Text-File Graph and Text-File Autostatistics

from Sensible Software, Bloomfield, MI, USA:
Sensible Speller, B.E.S.T. and all disk analysis programs for data analysis and safety

from Beagle Bros San Diego, CA, USA:
Alphaplot, Alphagraph

from Synergistic Software, Bellevue, WA, USA
Data Reporter

from A.P.P.L.E., Seattle, WA, USA:
Applesoft Carpenter, Apfile

This whole volume was written and produced with Apple Writer II from Paul Lutus (Apple Inc. Cupertino, CA, USA)

References

[1] *R.E. Kaiser, R.I. Rieder,* HRC&CC **2** (1979) 416; see also *V. Pretorius et al.* HRC&CC **1** (1978) 200.

[2] *P.K. Warme,* State College, USA, private communication to the authors (1979).

[3] *M.L. Peterson, J. Hirsch,* J. Lipid Res. **1** (1959), 132; see *H.J. Gold* Anal.Chem. **34** (1962) 174.

[4] *A.S. Said*: "Theory and Mathematics of Chromatography" (1981), Huethig, Heidelberg, Basel, New York, p. 148.

[5] *J. Sevcik,* private communication with the authors (1979).

[6] *S. Ebel, R.E. Kaiser,* Chromatographia **7** (1974) 696.

[7] *R.E. Kaiser, E. Oelrich*: "Optimization in HPLC" (1981), Huethig, Heidelberg, Basel, New York.

[8] *K. Bischoff,* in IfC HPLC training courses, private communication (1979...82).

[9] *M.J.E. Golay, J.G.Atwood,* J. Chromatogr. (1979), CHROM. 12,021 in Advances in Chrom. (1979) ed. *L.S. Ettre, E.sz. Kovats,* 369, and private communication of *M.J. Golay* with one of the authors (*REK*) 1979, 82.

[10] *Landau*: see *B. Schorr,* Comput.Phys.Commun. **7** (1974) 215; *P. Onuchin et al.,* NUCL **120(2)** (1974) 365.

[11] *W. Kimpenhaus, F. Richter, L. Rohrschneider,* Chromatographia **15** (1982) 577.

[12] *L.S. Ettre,* private communication (1982).

[13] *J. Krupcik, J. Garay, G. Guiochon, J.M. Schmitter,* Chromatographia **14** (1981) 501.

[14] *J.H. Knox* in *P.A. Bristow,* "LC in practice" (1976), p. 23, "hetp" Handforth, Wilmslow, Cheshire, UK.

[15] *A. Wehrli, E. Kovats,* Helv.Chimica Acta **42** (1959) 2709.

[16] *D. Post,* "Identification of Drugs...by Retention Indices" (in German), a book in this "Chromatographic Methods" series (1983), Huethig, Heidelberg, Basel, New York

[17] *R.E. Kaiser, R.I. Rieder, A. Miller, B. Pilgram* in "Proceedings of the Second Interntl. Symp. on HPTLC" (1982), 335, publ. by IfC, Bad Duerkheim (will be published in CAL—Computer Applications in the Laboratory, Huethig, Heidelberg, Basel, New York).

Selected Literature on Computer Use in the Laboratory

A.C. = ANALYTICAL CHEMISTRY

ANAL.CHIM.ACTA = ANALYTICA CHIMICA ACTA

CHR.= CHROMATOGRAPHIA (VIEWEG, GERMANY)

CHROM.TOKYO = ADVANCES IN CHROMATOGRAPHY CONFER. 1982 IN TOKYO (IN J.OF CHROMATOGRAPHY (1982))

HRC&CC = J. HIGH RESOLUTION CHROMATOGRAPHY AND CHROMATOGRAPHY COMMUNICATION (HUETHIG-NEW YORK)

PITTSBURGH CONF.= THE YEARLY PITTSBURGH CONFERENCE ABSTRACTS BOOK (ONLY THE 1982 ISSUE WAS SEARCHED)

THERE ARE ABOUT 80 OTHER JOURNALS DEALING WITH COMPUTER USE IN LABORATORIES NOT CITATED HERE

A COMP.ALGORITHM FOR THE REAL TIME SELECTION OF MASS SPECTR. D.HENNEBERG ET AL.,CHR.(1972),209

A COMP.SYSTEM FOR DATA ACQUISITION AND EVALUATION D.SCHEUTZOW,CHR.(1974),438

A COMPUTER-BASED COMPENDIUM FOR SELECT.INSTR.METH. F.A.SETTLE ET AL., PITTSBURGH CONF.

ADAPT.PROGRAM FOR HIGH PRECISION OFF-LINE PROCESS. M.H.J.VAN RIJSWICK,CHR.(1974),491

ALGORITHM FOR CONVERT.PEAK DETEC.AND TAILING INTO CHR.INFORMATION,W.A.SPENCER ET AL.,A.C.(1980),950

APPLICATION OF GC/FT-IR/MS TO ENVIRONM.POLLUT.SAMP ANALYS.,K.-H.SHAFER ET AL.,PITTSBURGH CONF.,1982

APPLICATION OF USER PROGRAMING WITH VISTA-PLUS F.KLINK ET AL.,PITTSBURGH CONF.,1982

CAN ANALOQUE METHODS SOLVE THE TRANSMISS.PROBLEMS?
G.A.SCHLERETH ET AL.,CHR.(1972),70

CHROM.DATA PROCESSOR WITH EXPAN.FUNCTION.CAPABIL.
T.SATO ET AL., PITTSBURGHH CONF.,1982

CHROMAT.RESOLUTION CONFIRMATION BY DIFFER.PLOTTING
M.S.TARTER ET AL.,PITTSBURGH CONF.,1982

COMPUTER SIMULATION OF THE CHROMATOGRAPHIC PROCESS
J.A.JONSSON,CHR.(1980),273

COMPUTER/CHROMATOGRAPH INTERFACE
H.BROWN ET AL.,PITTSBURGH CONF.,1982

COMPUTERS IN ANAL.CHEM.(WITH EMPHASIS ON CHROMAT.)
J.F.K.HUBER,CHR.(1974),421

DESIGN PHILOSOPHY FOR SEC.GENERATION LAB.COMPUTERS
R.J.KLIMOWSKI ET AL.,CHR.(1974),452

FRACTAL ANALYSIS APPLIED TO FT-IR INTERFEROGRAMS
D.F.LECLERC ET AL.,PITTSBURGH CONF.,1982

FTIR APPLICATION TO CHROMATOGRAPHY AND ANAL.PROBL.
J.M.CASPER ET AL., PITTSBURGH CONF., 1982

GPC SOFTWARE FOR AN INTELLIGENT GRAPHICS TERMINAL
R.L.MILLER,PITTSBURGH CONF.,1982

IMPROV.DATA HANDLING THROUGH THE PERS.COMP.INTERF.
D.J.KLOPP ET AL.,PITTSBURGH CONF.,1982

INTERFACING AUTOM.CHROMATOGRAPHS TO COMPUT.SYSTEMS
J.G.LYONS,CHR.(1972),156

LARGE COMP.SYST.FOR ON-LINE DATA ACQU.FROM ANAL.
INSTRUMENTS, FOK ET AL.,CHR.(1974),423

LES ERREURS EN CHROMATOGRAPHIE
R.E.KAISER, CHR.(1971),129

MICROPROCESSORS IN CHEMICAL INSTRUMENTATION
F.BAUMANN ET AL.,CHR.(1974),530

MULTIPLEXER OR DEDICATED ANALOG-TO-DIGIT.CONVERTER
U.BUSCH,CHR.(1972),63

OPTIMIZATION OF CHROM.SEPARATIONS BY COMP.CONTROL
S.P.CRAM ET AL.,CHR.(1974),567

POLYGON DIAGRAMS OF CHROMATOGRAPHIC DATA
G.JANSSENS, HRC&CC (1981),99

POWERFUL COMP.FOR ACQU. AND PROCESSING CHROM.AND
OTHER ANAL. DATA,GUICHARDET AL.,CHR.(1972),83

QUAL.U.QUANT.VERGL.VON CHROM.UNTER RECHNEREINSATZ
K.WASSERFALLEN ET AL.,CHR.(1977),176

QUANT.ANALYSIS OF NON-GAUSS.PEAKS BY COMP.CURVE FIT. A.H.ANDERSON ET AL.,CHR.(1969),466

QUANT.RESOLUTION OF SEVER.OVERLAPPING CHROM.PEAKS, J.T.LUNDEEN ET AL.,A.C.(1981),1369

REAL-TIME PEAK DETECTION AND AREA APPORTIONMENT A.FOZARD ET AL.,CHR.(1972),130

REPLOTTING,STORAGE AND RETRIEVAL OF CHROMAT.DATA T.F.MAYERS ET AL.,PITTSBURGH CONF.,1982

SIM.ANAL.OF TWO OR MORE SAMP.ON THE SAME CHR.COL. C.LAURGEAU ET AL.,CHR.(1979),160

STORAGE AND RETRIEVAL OF CHROM.DATA BY COMPUTER R.M.PETERSON ET AL.,CHR.(1972),146

THE REMOTE BATCH TERMINALAS AN EFF.FIRST STEP P.B.STOCKWELL ET AL.,CHR.(1974),465

THE USE OF NEW COMPUT.TECHNOLOGY IN CHROMATOGRAPHY L.MIKKELSEN ET AL.,CHR.(1974),447

TIME SHARING COMPUTER AND ITS ROLE IN CHROM.LAB. J.M.GILL ET AL.,CHR.(1972),108

TRENDS IN CHROMATOGRAPHIC AUTOMATION R.E.ANDERSON,CHR.(1972),105

VOLTAG.CONTR.OSCILLATOR FOR SIGNAL-PROP.DATA ACQUI SITION RATES,D.E.LEYDEN ET AL.,A.C.(1977),681

APPLICATION OF AN INEXP.,GENER.-PURP.MICROCOMPUTER M.GOEDERT ET AL.,CHR.(1974),539

A NOV.DETERM.OF PEAK AREAS FOR APPL.ON LARGE COMP. F.CAESAR ET AL.,CHR.(1974),526

A TOTAL SYSTEM FOR AUTOMATED GAS CHROMATOGRAPHY G.T.PAUL,CHR.(1972),192

AN EXAMPLE FOR COMPU.DATAPROCESSING FOR SMALL GC LABORATORIES, K.RATHGEB, CHR.(1971),270

APPL.OF SIMCA MULTIVAR.DATA ANAL.TO THE CLASSIF.OF GC PROFILES, S.W.WOLD ET AL.,AL.CHEM.ACTA (81),251

AUT.DATA PROCESSING IN CHROM.- A MIXED BLESSING P.B.STOCKWELL ET AL.,CHR.(1980),665

AUTOMATION OF GAS ANALYSIS WITH A COMPUT.GC SYSTEM G.CHARRIER ET AL.,CHR.(1972),119

AUTOMATISCHE DATENVERARBEITUNG IN DER GC J.BAUDISCH,CHR.(1969),443

COMP.ANALYSIS OF GAS CHROMATOGRAMS OF CONT.SAMP. MIXTRS,H.CLOUGH ET AL.,CHR.(1972),351

COMP.EVALUATION OF GAS CHROMATOGRAMS W.WIDE DYNAM.
R ET AL.,CHR.(1974),554

TOBACCO SMOKE,

COMPUTER COMPARISONS OF VARIABLES IN CAP.GC
D.F.INGRAHAM ET AL.,HRC&CC (1982),227

DATA PROCESSING OF GLC PESTICIDE ANALYSES
L.G.M.TH.TUINSTRA ET AL.,CHR.(1971),468

DESIGN AND CONSTRUCTION OF A COMP.-CONTR.GC INSTR.
A.SABATIER ET AL.,CHR.(1974), 560

DESIGN OF A GAS CHROMATOGRAPH FOR COMP.CONTR.OPERATION,J.M.WELLAND ET AL.,CHR.(1972),136

EFFECT OF COL.TEMP.CHANGES IN IDENTIF.USING GC DATA PROCESSING SYSTEMS,RIEDMANN,CHR.(1973),129

EFFECTS OF NORMALIZ.ON FEAT.SELECTION IN PYR.GC OF COAL TAR PITCHES,M.S.KLEE ET AL.,A.C.(1981),801

EXPERIENCES WITH A GC-COMPUT.SYSTEM IN AN IND.ANAL LABOR., J.BAUDISCH,CHR.(1972),79

ITER.BERECHNUNG DER TOTZEIT AUS BRUTTORETENTIONSD.
S.EBEL ET AL, CHR.(1974), 696

KOVATSINDICES ZUR IDENTIFIZ.VON PEAKS MIT GC COMP.
SYSTEMEN,F.CAESAR,CHR.(1972),173

LOGARITHMIC DATA COMPRESSION STRATEGY FOR GC
D.L.DOERFLER ET AL.,A.C.(1978),1709

METHOD AND PROGRAM FOR ACC.KOVATS RETENTION INDEX
I.B.PEETRE,CHR.(1973),257

ON-LINE TESTING OF GC COLUMNS WITH PROGRAMMABLE
COMP.INTEGRAT.,W.KIMPENHAUS ET AL.,CHR.(1982),577

PERFORMANCE APPR.OF HEWLETT PACKARD GC-COMP.SYSTEM
E.MALAN ET AL.,CHR.(1972),182

PLAUSIBILITY CHECKS IN COMPUTERIZED GC
J.BAUDISCH ET AL.,CHR.(1974),549

PPB-ANALYSEN DURCH COMPUTER ANWENDUNG?
R.KAISER,CHR.(1972),177

PRAG-EIN PROZESSRECHNER-AUSWERTESYSTEM FUER GC
H.D.METZGER ET AL.,CHR.(1972),186

PROBL.AND SOLUT.TO MEASURE AND MAINTAIN THE HIGH
SEP.EFF. OF CAP.COL.AT OPTIM.,KAISER,CHR.(1975),49

PROGRESS AND EXPERIENCES WITH MUELHEIM COMP.SYST.
IN GC DATA HANDL.,SCHOMBURG ET AL.,CHR.(1972),96

SOFTWARE PACKAGE TO COLLECT AND PROCESS RADIO GC DATA, I.M.CAMPBELL ET AL.,A.C.(1977),1726

SOFTWARE-PARAMETER OPTIMIZATION IN COMP.GC SYSTEM P.SUTRE ET AL.,CHR.(1972),141

TESTING,EVALUATION,COMPARISON,OPTIMIZATION OF CAP. GC.,R.E.KAISER ET AL.CHR. (1977),455

TIME-SHARING(APL)COMPUTERASSISTANCE IN GLC M.MARGOSIS,CHR.(1974),441

VERARBEITUNG GC DATEN DURCH EL.INTEGR.MIT COMP.SYS G.BISCHET ET AL., CHR.(1972), 166

COMPUTER RECONSTRUCTED GAS-AND LIQU.CHROMATOGRAMS G.JANSSENS, HRC&CC (1979),84

COMPUTER-AIDED MULTICHANNEL DETECTION IN HPLC A.F.FELL ET AL.,CHR.VOL.16 (1982),69

HPLC / VIDEO FLUOROMETRY M.P.FOGARTY ET AL.,HRC&CC (1981),561

MANIPULATION OF MULTI-FACET.DATA INCL.THE ANALYSIS OF LC CHROMATOGRAMS,M.FOX,CHR.(1974),510

MULTIWAVEL.DETECTION FOR LC WITH A REP.SCANN. UV-VIS SPECTROPHOTOMTR,K.SAITOH ET AL.,A.C.(1979),168

NEW GRAD.ELUT.AND TEMP.PROGR.METHODS BY A MICRO-PROC.CONTR.,A.WADA ET AL.CHROM.,TOKYO(1982)

OPTIMIZATION OF PHASE HPLC SEP.WITH A MICROCOMP. CONTR.CHROMATOGRAPH,J.C.BERRIDGE,CHR.VOL.16(1982

PERFORM.OF A NEW MICROPROOZESSOR CONTR.LC SYSTEM A.NAKAMOTO ET AL.,CHROM.TOKYO,1982

TWO-WAY DATA COMMUNICATION IN ROUTINE HPLC H.KNOESEL,CHR.(1979),362

AN AUTOMATED SYSTEM FOR CGC-IR ANALYSIS S.LOWRY ET AL.,PITTSBURGH CONF.,1982

AUT.COMPUTER USE OF MS AND RETENTION INDEX JOINT LIBRARIES, B.E.BLAISDELL, A.C.(1977),180

AUT.PRE-INTERPRETATION OFANAL.INF.OF GC/MS SYSTEMS J.FRANZEN,CHR.(1974),518

AUT.SIM.QUAL.AND QUANT.ANAL.OF COM.ORG.MIXTURES GC/MS-COMP.SYST.,S.C.GATES ET AL.,A.C.(1978),433

AUTOM.TUNING OF A BENCHTOP GC/MS F.CHEM.ION OPERATION,D.S.PEARCE ET AL.,PITTSBURGH CONF.,1982

CAPILLARY GC/FT-IR V.ROSSITER,PITTSBURGH CONF.,1982

CHROMATOGRAM SURGERY
G.JANSSENS, HRC&CC (1980),395

COMB.GC/IR/MS SEARCH SYST.USING DATA COMPR.BY FACT ANALYS.,S.S.WILLIAMS ET AL.,PITTSBURGH CONF.,1982

COMPAR.OF METHODS FOR RECONST.OF GAS CHROMATOGRAMS INT.GC/IRS DATA,R.L.WHITE ET AL.,A.C.(1981),1778

CROSS-CORREL.SIGNAL TO NOISE ENHANCEMENT WITH APPL TO QUANT.GC/FTIR,R.B.LAM ET AL.,PITTSB.CONF.1982

DATA-BLOCKING CROSS-CORREL.PEAK DETECTION IN COMP. GC-MS,WM.F.BRYANT ET AL.,A.C.(1980),38

DEVELOPMENT OF A GC/FTIR INTERFACE FOR REFR.SCANN. FTIR SPECTROMETER,J.N.SUN,PITTSBURGH CONF.,1982

DEVELOPMENTS IN CHROM.IR TECHNIQUES FOR THE ANAL. PESTICIDE,K.S.KALASINSKY ET AL.,PITTSB.CONF.1982

HIGH RESOLUTION CAPILLARY GC/FTIR
S.E.GARLOCK ET AL.,PITTSBURGH CONF.,1982

INTERAC.PROCESSING OF CHROMATOGRAMS WITH DISP.TERMINALS, G.SCHOMBURG ET AL., CHR.(1974),502

QUAN.ANAL.OF UNLABEL.AND POLYDEUT.COMPOUNDS BY GC-MS,M.AXELSON ET AL.,CHR.(1974),502

QUAN.GC/FTIR USING INTEG.GRAM-SCHMIDT RECONSTR.INTENSITY, D.T.SPARKS ET AL.,PITTSBURGH CONF.,1982

SOME ALGORITHM FOR THE COMP.EVALUATION OF CHROMATO GRAMS,B.WEIMANN,CHR.(1974),472

STAT.PREDICTION OF RESULTS FOR VAP.PHASE IR-IDENT. OF GC PEAKS,M.F.DELANEY ET AL.,A.C.(1979),1243

QUANTITATIVE EVALUATION OF THIN-LAYER CHROMATOGR.
M.PROSEK ET AL., HRC&CC (1981) 138

QUANTITATIVE EVALUATION OF THIN-LAYER CHROMATOGR.
S.LENARCIC ET AL.,HRC&CC (1982),694

VOLLAUT.AUSWERTUNG VON DC 2.MITT.:DIGITALSTEUERUNG
S.EBEL ET AL.,CHR.(1976),78

Polynomial Regression, a BASIC Program

```
10  REM   -----POLYN-E 8/82-REK

20  TEXT :
    LET A = 0:
    LET A$ =  CHR$ (4):
    TEXT :
    HOME :
    POKE 34,2:
    INVERSE :
    PRINT "***   POLYNOMIAL    INTERPOLATION    *4*":
    NORMAL :
    PRINT :
    HTAB 7:
    PRINT " (1.-6.DEGREE REGRESSION) ":
    PRINT :
    HTAB 4:
    FLASH :
    PRINT " <COPYRIGHT IFC BAD DUERKHEIM> ":
    NORMAL

30  FOR P = 1 TO 2000:
    NEXT P:
    VTAB 5:
    PRINT "FOR DATA ANALYSIS. DO NOT EXTRAPOLATE!!":
    PRINT :
    PRINT " X-DATA   (CONCENTRATION)"

40  PRINT " Y-DATA   (PEAK AREA OR SIGNAL ETC.)":
    FOR P = 1 TO 3500

50  NEXT P:
    VTAB 7:
    PRINT "                                        ":
    PRINT "                                        ":
    PRINT :
    PRINT :
    DIM A(19),B(19),C(11,11),D(50),E(50),F(50),C(50):
    DIM C$(50),D$(50),H(50),I(50)

60  INVERSE :
    PRINT "*  START WITH SMALLEST X  (SORTED!)    ":
    NORMAL

70  PRINT "":
    PRINT " ENTER DATA MANUALLY   (M)   OR ":
    PRINT " FROM DISK ?           (D)":
    HTAB 25:
    INPUT "";E$:
    IF E$ <  > "M" THEN
         GOSUB 1070

80  IF E$ = "D" THEN


     PRINT :
    PRINT :
    GOTO 160
```

```
 90  HOME :
     PRINT "NUMBER OF X-Y-DATA GROUPS (N MAX = 50)":
     PRINT :
     INPUT "                N = ";F$:
     LET F =  VAL (F$):
     PRINT :
     IF F = 0 THEN
         180

100  PRINT "CONC --> X,  SIGNAL --> Y  AVOID ERRORS":
     PRINT "---------------------------------------":
     POKE 34,12:
     FOR G = 1 TO F:
         VTAB 11:
         PRINT :
         PRINT "     X(";G")            Y(";G")       N=";G

110      LET H = 22

120      VTAB H:
         INPUT "     ";C$(G):
         IF C$(G) = "K" THEN
              LET G = G - 1:
             LET H = H - 1:
             GOTO 130

130      LET D(G) =  VAL (C$(G)):
         IF G >  = 1 AND D(G) = 0 THEN
              PRINT " X= ZERO OR (RETURN) NOT ACCEPTED  !!  ":
             FOR I = 1 TO 1500:
         NEXT I:
         GOTO 110

140      VTAB H:
         HTAB 20:
         INPUT "";D$(G):
         IF D$(G) = "K" THEN
             120

150      LET G(G) =  VAL (D$(G)):
         PRINT :
         PRINT :
     NEXT :
     TEXT

160  PRINT "*  USE (RETURN) TO SKIP QUESTIONS     ":
     PRINT "WANT (LOG) TRANSFORMATION (Y:N) ?":
     HTAB 38:
     INPUT "";E$:
     IF E$ = "Y" THEN


          GOSUB 790

170  LET A = 0:
     LET L = 0:
     LET N = 0:
     LET O = 0:
     LET P = 0:
```

```
     LET Q = 0:
     LET U = 0:
     PRINT :
     PRINT "POLYNOM M (1=LIN;2=SQR.;3=CUB.;MAX=6) ":
     INPUT "          ";M:
     PRINT :
     IF M > 6 THEN
         170


180  IF M = 0 THEN
          HOME :
         PRINT " * BYE *   RESTART BY '& RETURN'":
         END


190  IF M > F - 2 THEN
          PRINT "":
         INVERSE :
         PRINT "MATHEM. NOT ACCEPTABLE; M=<N-2!":
         NORMAL

200  IF M =  < F - 2 THEN
         220


210  PRINT :
     LET A = 0:
     LET L = 0:
     LET N = 0:
     LET O = 0:
     LET P = 0:
     LET Q = 0:
     LET U = 0:
     PRINT :
     PRINT "POLYNOM M (1=LIN;2=SQR;3=CUB.;MAX=6)":
     INPUT "          ";M:
     PRINT :
     IF M > 6 THEN
         170


220  PRINT :
     PRINT " ........PLEASE WAIT.........":
     FOR G = 1 TO 2 * M + 2:
         LET A(G) = 0:
         LET B(G) = 0:
     NEXT :
     FOR J = 1 TO M + 2:
         FOR T = 1 TO M + 2:
             LET C(T,J) = 0:


         NEXT :
     NEXT :
     LET A(1) = F:
     FOR G = 1 TO F:
         LET W = D(G):
         LET X = G(G)
```

```
230       FOR J = 2 TO 2 * M + 1:
              LET A(J) = A(J) + W ^ (J - 1):
          NEXT :
          FOR J = 1 TO M + 1:
              LET B(J) = B(J) + X * W ^ (J - 1):
              LET C(J,M + 2) = B(J):
          NEXT :
          LET B(M + 2) = B(M + 2) + X ^ 2:
      NEXT

240   FOR G = 1 TO M + 1:
          FOR J = 1 TO M + 1:
              LET C(G,J) = A(G + J - 1):
          NEXT :
      NEXT :
      FOR S = 1 TO M + 1:
          FOR T = S TO M + 1:
              IF C(T,S) <  > 0 THEN
                   GOTO 260

250       NEXT :
      NEXT :
      PRINT "DATA ERROR ??  ":
      PRINT :
      PRINT " RESTART BY '& RETURN' ":
      PRINT " AND REENTER THE (CORRECT) DATA":
      STOP

260   GOSUB 290:
      LET C = 1 / C(S,S):
      GOSUB 300:
      FOR T = 1 TO M + 1:
          IF T = S THEN
               GOTO 280

270       LET C =  - C(T,S):
          GOSUB 310

280   NEXT :
      NEXT :
      GOTO 320

290   FOR J = 1 TO M + 2:
          LET B = C(S,J):
          LET C(S,J) = C(T,J):
          LET C(T,J) = B:
      NEXT :


      RETURN

300   FOR J = 1 TO M + 2:
          LET C(S,J) = C * C(S,J):
      NEXT :
      RETURN

310   FOR J = 1 TO M + 2:
          LET C(T,J) = C(T,J) + C * C(S,J):
      NEXT J:
      RETURN
```

```
320  PRINT "-------------------------------------":
     PRINT "POLYNOM ";M;". DEGREE":
     PRINT :
     FOR G = 1 TO M + 1:
         PRINT "A";G - 1; TAB( 23);C(G,M + 2):
     NEXT :
     LET A = C(1,M + 2):
     LET L = C(2,M + 2):
     LET N = C(3,M + 2)

330  :
     LET O = C(4,M + 2):
     LET P = C(5,M + 2):
     LET Q = C(6,M + 2):
     LET U = C(7,M + 2):
     LET S = 0:
     FOR G = 2 TO M + 1:
         LET S = S + C(G,M + 2) * (B(G) - A(G) * B(1) / F):
     NEXT

340  LET T = B(M + 2) - B(1) ^ 2 / F:
     LET C = T - S:
     LET G = F - M - 1:
     IF G = 0 THEN
          INVERSE :
         PRINT :
         PRINT :
         PRINT " *  DEGR.OF POLYNOM TOO HIGH !!   * ":
         FOR G = 1 TO 2500:
     NEXT :
     NORMAL :
     GOTO 170

350  IF G =  <  - 1 THEN
          PRINT :
         PRINT :
         INVERSE :
         PRINT " * STATISTIC DATA NON VALID! *":
         NORMAL :
         PRINT :
         INVERSE :
         PRINT "  DECR.OF POLYNOM TOO HIGH    ":


         NORMAL

360  LET J = S / M:
     LET K = C / G:
     PRINT :
     IF K = 0 THEN
          LET K = 1E - 30

370  PRINT "F"; TAB( 23);J / K:
     PRINT :
     LET J = S / T:
     PRINT "COEFF.DETERMINATION"; TAB( 23);J:
     PRINT "COEFF.REGRESSION"; TAB( 23); SQR (J):
     LET B = J:
     LET Y = C / G:
     PRINT "STANDARD ERROR"; TAB( 23); SQR ( ABS (Y))
```

```
380  PRINT "-----------------------------------------"

390  INVERSE

400  PRINT "GRAPH (1) * PRINT (2) * OTH. POLYNOM (3)STATIST.ANA
     LYSIS  (4) * CALCULATIONS (5)":
     NORMAL :
     INPUT "CHOOSE BY (NR); END WITH (0)          ";D:
     TEXT :
     IF D = 0 THEN
          PRINT :
         PRINT " BYE!  RESTART BY '& RETURN'":
         END

410  TEXT :
     HOME :
     ON D GOTO 420,640,170,700,860,950


420  PRINT :
     PRINT "SCALING X AND Y AXIS: (1)    (RETURN)":
     PRINT "CHANGE SCALING :       (0)":
     VTAB 2:
     HTAB 24:
     INPUT "";A1:
     IF B1 = 0 OR C1 = 0 THEN
         440

430  IF A1 = 1 THEN
          GOTO 530


440  PRINT :
     FLASH :
     PRINT "*  START X AND Y AXIS WITH ZERO <<<   *":
     NORMAL :
     PRINT "   (PROGRAM ACCEPTS OTHER DATA,BUT... ":
     VTAB 22:
     PRINT "X-MIN,X-MAX,MARK INTERVL.;X-MAX = "; INT (100 * D(F


     )) / 100;"   ":
     PRINT "                                          "

450  VTAB 23:
     HTAB 1:
     INPUT "^";D1:
     VTAB 23:
     HTAB 7:
     INPUT "^";B1:
     VTAB 23:
     HTAB 13:
     INPUT "^";E1:
     PRINT :
     PRINT :
     IF E1 > B1 THEN
          PRINT "":
         PRINT "TICK MARK INTERVAL MUST BE SMALLER !!! ":
         GOTO 450
```

```
460  LET F1 = 1E33:
     LET G1 =  - F1:
     FOR G = 1 TO F:
         IF G(G) > F1 THEN
             480

470      LET F1 = G(G):
         LET H1 = G

480      IF G(G) < G1 THEN
             500

490      LET G1 = G(G):
         LET I1 = G

500  NEXT :
     VTAB 22:
     PRINT "Y-MIN = "; INT (100 * F1) / 100" ,Y-MAX = "; INT (1
     00 * G1) / 100" MARK INTERV."

510  VTAB 23:
     HTAB 1:
     INPUT "^";J1:
     VTAB 23:
     HTAB 12:
     INPUT "^";C1:
     VTAB 23:
     HTAB 24:
     INPUT "^";K1:
     IF J1 =  > C1 OR K1 > C1 THEN
          PRINT "":
         PRINT "CHECK Y-SCALING !!":
         FOR I = 1 TO 1000:
     NEXT I:
     GOTO 510


520  LET W -  FRE (0)

530  HGR :
     HCOLOR= 3:
     HPLOT 0,0 TO 279,0 TO 279,159 TO 0,159 TO 0,0:
     FOR G = D1 TO B1 STEP E1:
         LET Z = G:
         GOSUB 620:
         HPLOT W,155 TO W,159:
     NEXT :
     FOR G = J1 TO C1 STEP K1:
         LET V = G:
         GOSUB 630:
         HPLOT 0,X TO 4,X:
         HPLOT 275,X TO 279,X:
     NEXT :
     FOR G = 1 TO F:
         LET Z = D(G):
         GOSUB 620:
         LET V = G(G):
         GOSUB 630:
         IF W - 1 < 0 THEN
             GOTO 570
```

```
540      IF W + 1 > 279 THEN
              GOTO 570

550      IF X - 1 < 0 THEN
              GOTO 570

560      HOME :
         HPLOT W,X + 1 TO W,X - 1:
         HPLOT W - 1,X TO W + 1,X


570  NEXT :
     VTAB 23:
     PRINT "...PLEASE WAIT....":
     FOR R = 1 TO 279 STEP 5:
         LET G = (R - 1) * (B1 - D1) / 279:
         LET L1 = A + L * G + N * G ^ 2 + O * G ^ 3 + P * G ^ 4
          + Q * G ^ 5 + U * G ^ 6:
         LET M1 = 279 / B1:
         LET N1 = 159 / C1:
         LET O1 = G * M1:
         LET L1 = L1 * N1:
         IF O1 > 0 AND O1 < 279 AND L1 > 0 AND L1 < 159 THEN
              HPLOT O1,159 - L1


580  NEXT :
     LET Z = G:
     GOSUB 620:
     IF S < J1 THEN
          LET S = J1


590  IF S > C1 THEN
          LET S = C1

600  LET V = S:
     GOSUB 630:
     INVERSE :
     VTAB 22:
     GOTO 400

610  TEXT :
     HOME :
     GOTO 170


620  LET W = 279 * (Z - D1) / (B1 - D1):
     RETURN

630  LET X = 159 * (1 - (V - J1) / (C1 - J1)):
     RETURN

640  IF P1 = 0 AND Q1 = 0 THEN
          PRINT "FIRST CHECK QUALITY OF INTERPOL.DATA":
         PRINT " THEN CHOOSE   >>>  >>>        (2).....":
         FOR G = 1 TO 3500:
     NEXT :
     GOTO 700
```

```
650  LET A$ =  CHR$ (4):
     PRINT A$;"PR£1":
     PRINT :
     HTAB 6:
     PRINT " >>  S U M M A R Y  <<":
     PRINT :
     FOR G = 1 TO F:
         PRINT G;:
         HTAB 4:
         PRINT "X = ";D(G);:
         HTAB 19:
         PRINT "Y = ";G(G):
     NEXT

655  PRINT :
     PRINT "      B  = ";B:
     FOR G = 1 TO M + 1:
         PRINT "A";G - 1; TAB( 23);C(G,M + 2):
     NEXT :
     PRINT :
     PRINT "     POLYNOM  ";M;".DEGREE"

660  PRINT

670  PRINT "COEFF.OF CORREL.   = "; SQR (B):
     PRINT "STANDARD ERROR     = "; SQR ( ABS (C / G)):
     PRINT "QUALITY OF REGRESS.= "; INT (100 * R1) / 100" % +-

     "; INT (100 * S1) / 100" % "

675  IF  LEN (H$) > 0 THEN
          PRINT :
         PRINT "FILE NAME = ";H$

676  PRINT :
     PRINT :
     PRINT A$;"PR£0":
     PRINT :
     INVERSE

680  PRINT " USE SAME DATA? THEN (1); NEW DATA? (2)":
     NORMAL :
     INPUT " STATIST.DATA ANALYS.(3)              ";D:
     HOME :
     ON D GOTO 380,690,700

690  TEXT :
     HOME :
     CLEAR :
     GOTO 20

700  VTAB 10:
     HTAB 10:
     PRINT " .....PLEASE WAIT......":
     LET P1 = 0:
     LET Q1 = 0:
     FOR G = 1 TO F:
         IF G(G) = 0 THEN
              LET G(G) = 1E - 10:
     NEXT
```

```
710  FOR G = 1 TO F:
         LET E(G) = A + L * D(G) + N * D(G) ^ 2 + O * D(G) ^ 3
          + P * D(G) ^ 4 + Q * D(G) ^ 5 + U * D(G) ^ 6:
     NEXT :
     IF G$ = "LOGXY" THEN
          GOSUB 960

720  IF G$ = "K" THEN
          GOSUB 970



730  FOR G = 1 TO F:
         LET F(G) = 100 * ( ABS (E(G) - G(G)) / G(G)):
         LET P1 = P1 + F(G):
         LET Q1 = Q1 + F(G) ^ 2:
     NEXT :
     LET H = 22:
     TEXT :
     HOME :
     PRINT :
     PRINT "       **  S U M M A R Y  **  ":
     PRINT :



     PRINT "N  X     Y(MEAS)   Y(CALC.)  DIFF.(%)":
     PRINT "--------------------------------------":
     POKE 34,5

740  FOR G = 1 TO F:
         VTAB H:
         PRINT G:
         VTAB H:
         HTAB 4:
         PRINT  INT (1000 * D(G)) / 1000:
         VTAB H:
         HTAB 12:
         PRINT  INT (1000 * G(G)) / 1000:
         VTAB H:
         HTAB 24:
         PRINT  INT (10 * E(G) + .5) / 10

750      VTAB H:
         HTAB 35:
         PRINT  INT (10000 *  ABS ((G(G) - E(G)) / G(G)) + .5)
          / 100:
         PRINT :
         PRINT :
     NEXT :
     LET S1 =  SQR ( ABS (Q1 - P1 ^ 2 / F) / (F - 1)):
     LET S1 =  INT (1000 * S1 + .5) / 1000:
     LET R1 = P1 / F

760  PRINT " N = ";F" ;DEGR.POLYN. = ";M" ;B = "; INT (B * 1000
     000) / 1000000:
     HTAB 7:
     PRINT " REGRESSION QUALITY 'RQ':":
     INVERSE :
     HTAB 7
```

```
770  PRINT " 'RQ' = "; INT (100 * R1) / 100" % +- "; INT (100
      * S1) / 100" % ":
     NORMAL :
     HTAB 32:
     INPUT "(RETURN)";E$:
     IF E$ = "N" THEN
          END
780  PRINT :
     GOTO 390

790  PRINT " X-->LOG(X)? Y-->LOG(Y)? ENTER (X);(Y) ":
     PRINT :
     VTAB 23:
     HTAB 10:
     INPUT "";C$:
     VTAB 23:
     HTAB 22:
     INPUT "";D$:
     IF C$ = "X" AND D$ = "Y" THEN
         850
800  IF C$ = "X" THEN
         830

810  IF D$ = "Y" THEN
         840

820  IF C$ <  > "X" THEN
          RETURN

830  FOR G = 1 TO F:
         LET D(G) =  LOG (D(G)) /  LOG (10):
     NEXT :
     LET G$ = "LOGX":
     RETURN

840  FOR G = 1 TO F:
         LET G(G) =  LOG (G(G)) /  LOG (10):
     NEXT :
     LET G$ = "LOGY":
     RETURN

850  FOR G = 1 TO F:
         LET D(G) =  LOG (D(G)) /  LOG (10):
         LET G(G) =  LOG (G(G)) /  LOG (10):
     NEXT :
     LET G$ = "LOGXY":
     RETURN

860  TEXT :
     HOME :
     POKE 34,3:
     INVERSE :
     PRINT ">>> INTERPOL.     X(BASIS)   ;Y(CORREL) ":
     NORMAL :
     IF  LEN (G$) <  > 0 THEN
          INVERSE :
         PRINT "=> NOTE: LOG DATA TRANSFORMATION !!  <=":
         NORMAL
```

```
870  PRINT "STORE DATA TO DISK (Y); END PROGRAM (/)":
     PRINT "---------------------------------------":
     POKE 34,4

880  PRINT :
     INPUT " X = ";C$:
     LET W =  VAL (C$):
     IF (W > D(F) OR W < D(1) AND C$ <  > "/") THEN
          PRINT "* NOTE: DATA OUTSIDE CALCULATED RANGE! ":
         PRINT ""

890  IF G$ = "LOGXY" AND W = 0 THEN

          LET W = 1E - 20

900  IF (G$ = "LOGXY" OR G$ = "LOGX") THEN
          LET W =  LOG (W) /  LOG (10)

910  IF C$ = "/" THEN
          GOTO 390

920  IF C$ = "Y" THEN
         980

930  LET X = A + L * W + N * W ^ 2 + O * W ^ 3 + P * W ^ 4 + Q
      * W ^ 5 + U * W ^ 6:
     LET X =  INT (10000 * X + .5) / 10000:
     IF (G$ = "LOGXY" OR G$ = "LOGY") THEN
          LET X = 10 ^ X

940  PRINT "            Y = ";X:
     GOTO 880

950  GOTO 400

960  FOR G = 1 TO F:
         LET E(G) = 10 ^ E(G):
         LET G(G) = 10 ^ G(G):
     NEXT :
     RETURN

970  FOR G = 1 TO F:
         LET E(G) = 1 / E(G):
         LET G(G) = 1 / G(G):
     NEXT :
     RETURN

980  TEXT :
     HOME :
     PRINT " WANT TO STORE RAW DATA TO DISK?   (RD)":
     PRINT :
     PRINT " WANT TO CHANGE X AGAINST BASIS Y ?     ":
     PRINT " (X=CORRELATED TO Y!!)   ":
     HTAB 36:
     PRINT "(YX)":
     PRINT :
     HTAB 38:
     INPUT "";E$:
     IF E$ = "RD" THEN
         1030
```

```
 990  IF E$ = "YX" THEN
          1020

1000  IF E$ = "/" THEN
          870

1010  GOTO 980

1020  FOR G = 1 TO F:
          LET H(G) = D(G):
          LET I(G) = G(G):
          LET G(G) = H(G):
          LET D(G) = I(G):
      NEXT :
      HOME :
      PRINT "CHANGING Y->X TO X->Y   :":
      PRINT

1025  PRINT "N          NOW = X          NOW = Y":
      PRINT "---------------------------------------":
      FOR G = 1 TO F:
          PRINT G,D(G),G(G):
      NEXT :
      PRINT :
      PRINT :
      GOTO 160

1030  PRINT :
      PRINT "  DATA FILE NAME     = ":
      LET A$ =  CHR$ (4):
      PRINT "  ('?' FOR CATALOG)";:
      HTAB 26:
      INPUT "";H$:
      IF H$ = "?" THEN
           PRINT A$"CATALOG"

1040  IF H$ = "?" THEN
           PRINT :
          INVERSE :
          INPUT " FILE NAME = ";H$:
          NORMAL

1050  PRINT A$;"OPEN ";H$:
      PRINT A$;"WRITE ";H$:
      PRINT F:
      FOR G = 1 TO F:
          PRINT D(G):
          PRINT G(G):
      NEXT :
      PRINT A$;"CLOSE ";H$:
      PRINT :
      PRINT " END (/) ; CONTINUE (RETURN)":
      HTAB 25:
      INPUT "";E$:
      IF E$ = "/" THEN
           END

1060  GOTO 160

1070  PRINT :
```

```
        PRINT " (?) FOR CATALOG":
        INPUT " FILE NAME :     ";H$:
        IF H$ = "?" THEN
             PRINT A$"CATALOG":
            GOTO 1070

1080    REM

1090    PRINT :
        PRINT "N              X(I)              Y(I)":
        PRINT "------------------------------------":
        PRINT A$;"OPEN ";H$:
        PRINT A$;"READ ";H$:
        INPUT F:
        IF F > 50 THEN
             LET F = 50

1100    FOR G = 1 TO F:
            INPUT D(G),G(G):
            PRINT G;:
            HTAB 15:
            PRINT D(G);:
            HTAB 30:
            PRINT G(G):
        NEXT :
        PRINT A$"CLOSE ";H$:
        RETURN

1110    END

1120    REM        COPYRIGHT IFC *  INSTITUTE FOR CHROMATOGRAPHY *
          P.O.BOX 1141 *          D-6702 BAD DUERKHEIM-1      WES
        T GERMANY                DR.R.E.KAISER

END-OF-LISTING
```

Index

Hüthig

A Bonanza of Ideas for Chromatographers

R. E. Kaiser (Germany) (Editor)

Fourth International Symposium Hindelang 1981

1981, 944 pp., soft cover, $ 39.50
ISBN 3-7785-0743-5

This book contains the proceedings of the 4th International Symposium on Capillary Chromatography held in Hindelang/Germany in May 1981. The object of the symposium was to bring together leading scientists in high resolution chromatography from around the world in an intensiv 3-day meeting. As in the previous meetings in which many milestones in the development of chromatography have been set, this symposium produced an extraordinary number of exciting papers and novel ideals. The organizers of this symposium have strived to communicate state of the art science, technology and applications in capillary chromatography as an affordable book with a minimum of delay (less than 6 months).

The book is a bonanza of ideas and truly represents current developments in both gas and liquid chromatography with capillary columns. Specialists from 26 countries contributed 45 papers in all areas whitch touch upon the high resolution aspects of chromatography. Several papers deal with critical evalutions of glass surfaces, including fused silica and vitreous carbon. Approximately one half of the contribution cover applications from areas such as biochemistry, drug metabolism and environmental analysis. Specialists in the field of high resolution chromatography and related disciplines will find this book extremely helpful and informative. It also will be of interest to those concerned with analytical methodology, instrumentation and data handling. A few papers deal with theoretical aspects but most contributions aim toward the solution of practical problems. The book contains a large number of tables and figures and is effectively indexed. Although some discussions are presented at a high level, it is easily readable and can be appreciated by anyone with a background in chromatography.

Dr. Alfred Huethig
Publishers
Room 227
611 Broadway
New York, N.Y. 10012

Dr. Alfred Hüthig
Verlag GmbH
Postfach 102869
6900 Heidelberg 1

A-HRC-7

min
1
1009   60 min/hr
409   16
49

determine Peak height:

Polynomial < Regression / interpolation (noisy signal)

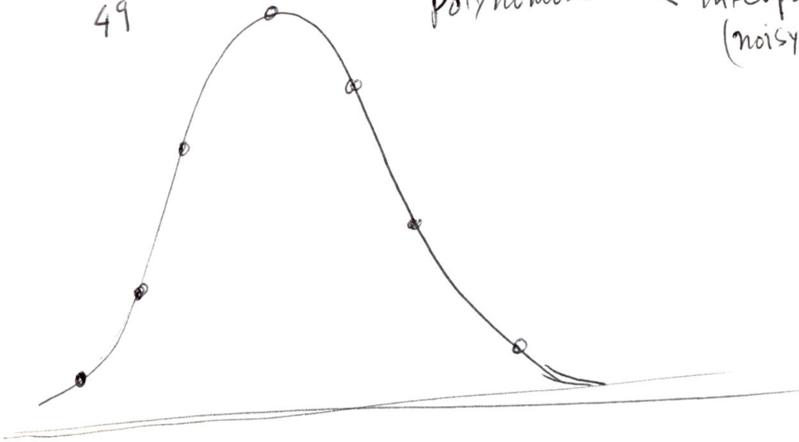

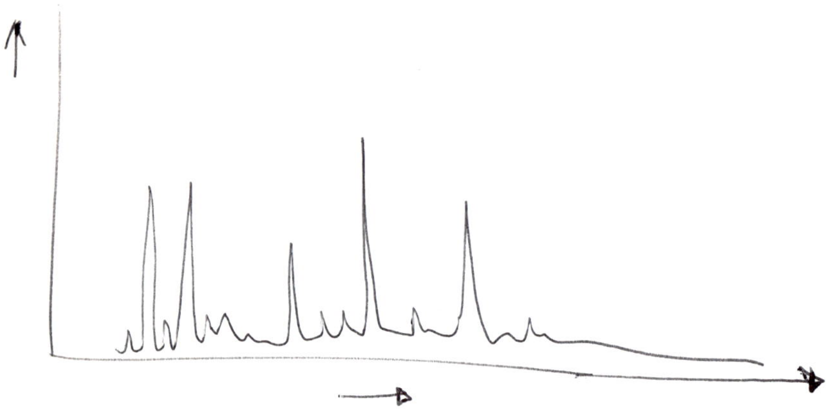